TASC Math Prep

The Ultimate Step-by-Step Guide Plus Two Full-Length TASC Practice Tests

Michael Smith
www.mathnotion.com

TASC Math Prep

Published By: The Math Notion

Web: www.mathnotion.com

Email: info@mathnotion.com

Copyright © 2021 by the Math Notion. All rights reserved. No part of this publication may be reproduced, stored in a retrieval system, or transmitted in any form or by any means, electronic, mechanical, photocopying, recording, scanning, or otherwise, except as permitted under Section 107 or 108 of the 1976 United States Copyright Ac, without permission of the author.

All inquiries should be addressed to the Math Notion.

ISBN: 978-1-63620-195-5

The Math Notion

Michael Smith has been a math instructor for over a decade now. He launched the Math Notion. Since 2006, we have devoted our time to both teaching and developing exceptional math learning materials. As a test prep company, we have worked with thousands of students. We have used the feedback of our students to develop a unique study program that can be used by students to drastically improve their math scores fast and effectively. We have more than a thousand Math learning books including:

- **HiSET Math Prep**
- **TABE Math Prep**
- **TASC Math Prep**
- **Accuplacer Math Prep**
- **Common Core Math Prep**
- **many Math Education Workbooks, Study Guides, Practice and Exercise Books**

As an experienced Math test preparation company, we have helped many students raise their standardized test scores—and attend the colleges of their dreams: We tutor online and in person, we teach students in large groups, and we provide training materials and textbooks through our website and through Amazon.

You can contact us via email at:

info@mathnotion.com

How to Achieve a Perfect Score on the TASC Math Test?

TASC Math Prep covers all mathematics topics that will be key to succeeding on the TASC math test. The step-by-step guide and hundreds of examples in this book can help you hone your math skills, boost your confidence, and be well prepared for the TASC test.

This new TASC self-teaching prep book offers extensive preparation and brush-up in math for those test-takers who plan to take the TASC Test. Two TASC math practice exams with detailed answers reflect questions and question types found on the actual test.

Updated the new TASC prep for 2020, 2021, and beyond by top test prep experts. Inside the TASC math preparation, You'll Find:

- Comprehensive TASC math review of key concepts
- Content 100% aligned with the latest TASC math exam
- Over 2,000 practice questions to help you master each TASC math topic
- Lots of examples with step-by-step solutions to illustrate all the TASC question types
- Each lesson is short, concise, and to the point.
- Focus on the most challenging part of the TASC math test!
- 2 full-length practice exams (featuring new question types) with detailed answers
- And much more than we can fit in this space…

After completing this math preparation book, you will discover your strengths and weaknesses, a strong foundation, and gain confidence to be successful on the TASC test. We have helped hundreds of thousands of people pass the TASC test and achieve their education and career goals. Get math prep for the TASC review that you need to ace your exam.

It is an excellent investment in your future!

www.MathNotion.com

... So Much More Online!

✓ FREE Math Lessons

✓ More Math Learning Books!

✓ Mathematics Worksheets

✓ Online Math Tutors

✓ For a PDF Version of This Book

Please Visit www.mathnotion.com

Contents

Chapter 1 : Whole Numbers, Real Numbers, and Integers 11
    Rounding 13
    Estimates 15
    Whole Number Addition and Subtraction 17
    Whole Number Multiplication 19
    Whole Number Division 21
    Adding and Subtracting Integers 23
    Multiplying and Dividing Integers 25
    Arrange, Order, and Comparing Integers 27
    Compare Integer 29
    Order of Operations 31
    Integers and Absolute Value 33

Chapter 2 : Fractions and Decimals 35
    Simplifying Fractions 37
    Factoring Numbers 39
    Greatest Common Factor (GCF) 41
    Least Common Multiple (LCM) 43
    Divisibility Rules 45
    Adding and Subtracting Fractions 47
    Multiplying and Dividing Fractions 49
    Adding Mixed Numbers 51
    Subtracting Mixed Numbers 53
    Multiplying Mixed Numbers 55
    Dividing Mixed Numbers 57
    Comparing Decimals 59
    Rounding Decimals 61
    Adding and Subtracting Decimals 63
    Multiplying Decimals 65
    Dividing Decimals 67
    Converting Between Fractions, Decimals and Mixed Numbers 69

Chapter 3 : Proportion, Ratio, Percent 71
    Writing Ratios 73
    Simplifying Ratios 75
    Create a Proportion 77
    Similar Figures 79
    Ratio and Rates Word Problems 81
    Percentage Calculations 83
    Percent Problems 85
    Markup, Discount, and Tax 87
    Simple Interest 89
    Converting Between Percent, Fractions, and Decimals 91

Chapter 4 : Exponents and Radicals 93
    Multiplication Property of Exponents 95
    Division Property of Exponents 97
    Powers of Products and Quotients 99

Zero and Negative Exponents 101
Negative Exponents and Negative Bases 103
Writing Scientific Notation 105
Square Roots 107

Chapter 5 : Algebraic Expressions 109

Translate Phrases into an Algebraic Statement.. 111
The Distributive Property 113
Evaluating One Variable 115
Evaluating Two Variables.. 117
Expressions and Variables 119
Combining like Terms 121
Expressions 121
Simplifying Polynomial Expressions 123

Chapter 6 : Equations and Inequalities 125

One–Step Equations 127
Two–Step Equations 129
Multi–Step Equations 131
Graphing Single–Variable Inequalities 133
One–Step Inequalities 135
Two–Step Inequalities 137
Multi–Step Inequalities 139
Solving Systems of Equations by Substitution.. 141
Solving Systems of Equations by Elimination 143
Systems of Equations Word Problems 145
Linear Equations 147
Graphing Lines of Equations 149
Graphing Linear Inequalities... 151
Finding Distance of Two Points 153

Chapter 7 : Polynomials 155

Classifying Polynomials.. 157
Adding and Subtracting Polynomials 159
Multiply and Divide Monomials 161
Multiplying Monomials... 163
Multiply a Polynomial and a Monomial 165
Multiply Binomials.. 167
Factor Trinomials 169
Operations with Polynomials 171
Simplifying Polynomials 173

Chapter 8 : Functions 175

Relations and Functions 177
Rate of change 179
Slope 181
x and y intercept... 183
Writing Linear Equations 185
Slope-intercept form 187
Point-slope form 189
Equation of Parallel or Perpendicular lines... 191
Equation of Horizontal and Vertical Lines 193
Function Notation.. 195
Adding and Subtracting Functions.. 197

Multiplying and Dividing Functions 199
Composition of Functions 201
Solve a Quadratic Equation 203

Chapter 9 : Geometry 205

The Pythagorean Theorem... 207
Angles 209
Area of Triangles 211
Area of Trapezoids 213
Area and Perimeter of Polygons 215
Area and Circumference of Circles 217
Volume of Cubes.. 219
Volume of Rectangle Prisms 221
Surface Area of Cubes... 223
Surface Area of a Rectangle Prism 225
Volume of a Cylinder.. 227
Surface Area of a Cylinder... 229

Chapter 10 : Statistics 231

Mean and Median 233
Mode and Range 235
Times Series 237
Box and Whisker Plot... 239
Bar Graph 241
Dot plots... 243
Scatter Plots 245
Stem–And–Leaf Plot 247
The Pie Graph or Circle Graph 249
Probability of Simple Events... 251

Chapter 11 : TASC Test Review 253

TASC Practice Tests Answer Sheet 255
TASC Mathematics Reference Sheet 257
TASC Practice Test 1 259
    Section 1 - Calculator... 259
    Section 2 - No Calculator 269
TASC Practice Test 2 273
    Section 1 - Calculator... 273
    Section 2 - No Calculator 283

Chapter 12 : Answers and Explanations 287

Answer Key.. 287
Practice Test 1 289
    Section 1- Calculator.. 289
    Section 2 - No Calculator 296
Practice Test 2 299
    Section 1 - Calculator... 299
    Section 2- No Calculator 306

Multiplying and Dividing Functions 199
Composition of Functions 201
Solve a Quadratic Equation 203

Chapter 9 : Geometry 205

The Pythagorean Theorem.. 207
Angles 209
Area of Triangles 211
Area of Trapezoids 213
Area and Perimeter of Polygons 215
Area and Circumference of Circles 217
Volume of Cubes.. 219
Volume of Rectangle Prisms 221
Surface Area of Cubes.. 223
Surface Area of a Rectangle Prism.. 225
Volume of a Cylinder.. 227
Surface Area of a Cylinder.. 229

Chapter 10 : Statistics 231

Mean and Median 233
Mode and Range 235
Times Series 237
Box and Whisker Plot... 239
Bar Graph 241
Dot plots.. 243
Scatter Plots 245
Stem-And-Leaf Plot 247
The Pie Graph or Circle Graph 249
Probability of Simple Events.. 251

Chapter 11 : TASC Test Review 253

TASC Practice Tests Answer Sheet 255
TASC Mathematics Reference Sheet 257
TASC Practice Test 1 259
    Section 1 - Calculator 259
    Section 2 - No Calculator 269
TASC Practice Test 2 273
    Section 1 - Calculator 273
    Section 2 - No Calculator 283

Chapter 12 : Answers and Explanations 287

Answer Key.. 287
Practice Test 1 289
    Section 1- Calculator 289
    Section 2 - No Calculator 296
Practice Test 2 299
    Section 1 - Calculator 299
    Section 2- No Calculator 306

Chapter 1 : Whole Numbers, Real Numbers, and Integers

Topics that you'll learn in this chapter:

- Rounding and Estimates
- Addition, Subtraction, Multiplication and Division Whole Number and Integers
- Arrange and ordering Integers and Numbers
- Comparing Integers, Order of Operations
- Mixed Integer Computations
- Integers and Absolute Value

"If people do not believe that mathematics is simple, it is only because they do not realize how complicated life is." — *John von Neumann*

Rounding

Rounding is replacing a number up or down to the closest number or the closest hundred, etc.
- ✓ First, you have to know the place value you'll round to.
- ✓ Second, you have to find the digit to the right of the place value you're rounding to. If it is 5 or greater, add 1 to the place value you're rounding to and put zero for all digits on its right side. If the digit to the right of the place value is smaller than 5 then keep the place value and put zero for all digits to the right.

EXAMPLE:

Round 64 to the closest ten.

The place value of ten is 6. The digit on the right side is 4 (which is smaller than 5). Now keep 6 and put zero for the digit on the right side. Now our answer is 60. 64 is rounded to the closest ten is 60, because 64 is closer to 60 than to 70.

PRACTICES:

Round each number to the underlined place value.

1) <u>8</u>8	2) <u>8</u>.15
3) <u>4</u>,315	4) 5<u>6</u>5
5) 1.<u>3</u>31	6) 14.<u>2</u>3
7) <u>2</u>.429	8) 4.3<u>1</u>3
9) 2.<u>9</u>97	10) <u>7</u>.38

Score: ....

Answer Key

1) 90	2) 8.0
3) 4,000	4) 570
5) 1.3	6) 14.2
7) 2.0	8) 4.31
9) 3.0	10) 7.0

Estimates

Estimating is a math policy used for approximating a number. To estimate means to make an irregular guess or calculation. To round means to make easier a known number by scaling it a little bit up or down.

- ✓ To estimate a math problem, round the numbers.
- ✓ For 2-digit numbers, you can usually round to the nearest tens, for 3-digit numbers, round to nearest hundreds, etc.
- ✓ Find the answer.

EXAMPLE:

Estimate the sum by rounding every number to the closest hundred. $153 + 426 = ?$

153 is rounded to the closest hundred which is 200. Now 426 is rounded to the closest hundred which is 400.

Then: $200 + 400 = 600$

PRACTICES:

Estimate the sum by rounding each added to the nearest ten.

1) 17 + 18	2) 94 + 81
3) 203 + 56	4) 55 + 33
5) 96 + 49	6) 99 + 324
7) 823 + 488	8) 466 + 276
9) 5,112 + 5,792	10) 1,245 + 2,459

Score: ....

Answer Key

1) 40	2) 200
3) 260	4) 90
5) 150	6) 400
7) 1,300	8) 800
9) 11,000	10) 3,000

Whole Number Addition and Subtraction

- ✓ Arrange the numbers in line.
- ✓ Start with the unit place. (Ones place)
- ✓ Regroup if needed.
- ✓ Add or subtract the tens place.
- ✓ Continue with further digits.

EXAMPLE:

Find the sum. $285 + 145 = ?$

First line up the numbers: $\begin{array}{r}285\\+145\\\hline\end{array}$ → Start with the unit place. (ones place) $5 + 5 = 10$,

Write 0 for ones place and keep 1, $\begin{array}{r}1\\285\\+145\\\hline 0\end{array}$, Add the tens place and the digit 1 we kept:

$1 + 8 + 4 = 13$, Write 3 and keep 1, $\begin{array}{r}1\,1\\285\\+145\\\hline 30\end{array}$

Continue with further digits → $1 + 2 + 1 = 4$ → $\begin{array}{r}1\,1\\285\\+145\\\hline 430\end{array}$

Find the difference. $976 - 453 = ?$

First line up the numbers: $\begin{array}{r}976\\-453\\\hline\end{array}$, → Start with the unit place. $6 - 3 = 3$, $\begin{array}{r}976\\-453\\\hline 3\end{array}$,

Subtract the tens place. $7 - 5 = 2$, $\begin{array}{r}976\\-453\\\hline 23\end{array}$, Continue with further digits → $9 - 4 = 5$,

$\begin{array}{r}976\\-453\\\hline 523\end{array}$

PRACTICES:

Find the missing number.

1) $540 - \ldots = 100$	2) $800 - \ldots = 220$
3) $\ldots - 2{,}650 = 6{,}700$	4) $85{,}000 - 42{,}000 = \ldots$
5) $1{,}280 - \ldots = 420$	6) $5{,}000 + 8{,}450 = \ldots$
7) $\ldots - 3{,}870 = 9{,}630$	8) $12{,}310 - \ldots = 8{,}540$

TASC Math Prep

Solve.

9) A school had 708 students last year. If all last year students and 218 new students have registered for this year, how many students will there be in total?

10) Lisa had $856 dollars in her saving account. She gave $295 dollars to her brother, Tom. How much money does she have left?

Score: ....

Answer Key

1) 440	2) 580
3) 9,350	4) 43,000
5) 860	6) 13,450
7) 13,500	8) 3,770
9) 926	10) 561

www.mathnotion.com

Name: ....

Whole Number Multiplication

- ✓ First you have to learn the times tables! To solve multiplication problems quick, you need to learn the times table. For example, 3 times 8 is 24 or 8 times 7 is 56.
- ✓ For multiplication, line up the numbers that you are multiplying.
- ✓ Start with the ones place and regroup if needed.
- ✓ Continue with further digits.

EXAMPLE:

Solve. $500 \times 30 = ?$

Line up the numbers: $\begin{array}{r} 500 \\ \times 30 \\ \hline \end{array}$, start with the ones place → $0 \times 500 = 0$, $\begin{array}{r} 500 \\ \times 30 \\ \hline 0 \end{array}$, Continue with further digit which is 3. → $3 \times 500 = 1{,}500$, $\begin{array}{r} 500 \\ \times 30 \\ \hline 15{,}000 \end{array}$

PRACTICES:

Multiply the Number.

1) $120 \times 6 = $ _____

2) $160 \times 30 = $ _____

3) $600 \times 30 = $ _____

4) $420 \times 20 = $ _____

5) $250 \times 40 = $ _____

6) $600 \times 40 = $ _____

7) $215 \times 70 = $ _____

8) $540 \times 11 = $ _____

9) $121 \times 10 = $ _____

10) $254 \times 16 = $ _____

Score: ....

Answer Key

1) 720	2) 4,800
3) 18,000	4) 8,400
5) 10,000	6) 24,000
7) 15,050	8) 5,940
9) 1,210	10) 4,064

Whole Number Division

- Division: A typical division problem: Dividend ÷ Divisor = Quotient
- In division, we want to find how many times a divisor is contained in a dividend. The result we obtain in a division problem is called quotient.
- First, the problem is written in division format. (Dividend is inside; divisor is outside)

Divisor | Dividend (Quotient on top)

EXAMPLE:

Solve. $234 \div 4 = ?$

First, write the problem in division format.

Start from left digit of the dividend. 4 won't divide 2.

So, we have to choose another digit of the dividend. It is 3.

Now, we will find how many times 4 goes into 23 and the answer is 5.

Write 5 above the dividend part. 4 times 5 is 20.

Write 20 below 23 and subtract. We get the answer 3.

Now take down the next digit which is 4 and find how many times 4 goes into 34?

The answer is 8. Write 8 above dividend.

This is last step since there is no further digit left.

of the dividend to bring down.

The final answer is 58 and we have the remainder 2.

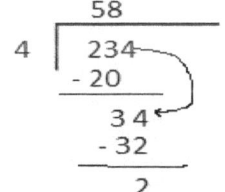

PRACTICES:

Divide the Number.

1) $450 \div 5 = $ _____	2) $320 \div 8 = $ _____
3) $125 \div 25 = $ _____	4) $720 \div 12 = $ _____
5) $588 \div 14 = $ _____	6) $299 \div 13 = $ _____
7) $869 \div 11 = $ _____	8) $801 \div 9 = $ _____
9) $493 \div 17 = $ _____	10) $600 \div 24 = $ _____

TASC Math Prep

Score: ....

Answer Key

1) 90	2) 40
3) 5	4) 60
5) 42	6) 23
7) 79	8) 89
9) 29	10) 25

Adding and Subtracting Integers

- ✓ Integers include zero, positive natural numbers, and the negative of the natural numbers. $\{\ldots, -3, -2, -1, 0, 1, 2, 3, \ldots\}$
- ✓ Add a positive integer by putting it to the right on the number line.
- ✓ Add a negative integer by putting it to the left on the number line.
- ✓ Subtract an integer by adding its opposite.

EXAMPLE:

Solve. $(-8) - (-5) =$

We keep the first number and change the sign of the second number to its opposite.

(Change subtraction into addition. Then: $(-8) + 5 = -3$

Solve. $10 + (4 - 8) =$

First subtract the numbers in brackets, $4 - 8 = -4$

Then: $10 + (-4) = \rightarrow$ changes addition into subtraction: $10 - 4 = 6$

PRACTICES:

Find the sum and difference.

1) $8 + (-11)$	2) $(-13) + 25$
3) $(55) - (21)$	4) $(4) - (-5) - (-3)$
5) $2 + (-11) + (-30) + (9)$	6) $(-5) + (-10) + (7 - 19)$
7) $(-20) - (-44)$	8) $(-9) - 13 + 20$
9) $(50) - (-5) + (-25)$	10) $24 + 16 + (-13)$

Score: ....

Answer Key

1) −3	2) 12
3) 34	4) 12
5) −30	6) −27
7) 24	8) −2
9) 30	10) 27

Multiplying and Dividing Integers

- ✓ (positive) × (positive) = positive
- ✓ (positive) ÷ (positive) = positive
- ✓ (negative) × (negative) = positive
- ✓ (negative) ÷ (negative) = positive
- ✓ (negative) × (positive) = negative
- ✓ (negative) ÷ (positive) = negative
- ✓ (positive) × (negative) = negative
- ✓ (positive) ÷ (negative) = negative

÷ / ×	+	−
+	+	−
−	−	+

EXAMPLE:

$(+5) \times (+3) = 5 + 5 + 5 = 15$

The basic idea of multiplication is recurrent addition. Example: 5 × 3 = 5 + 5 + 5 = 15

We know that division is the inverse operation of multiplication. So, 15÷3=5 because 5×3=15 In words, this expression says that 15 may be divided into 3 groups of 5 every because adding five thrice gives 15.

Divide (-91) by (-7)?

Examples on division of integers on different kinds of problems on integers are mentioned here step by step. (-91) ÷ (-7) =13

PRACTICES:

Find each product and each quotient.

1) (− 8) × (− 5)	2) 72 ÷ 9
3) 4 × (− 5) × (− 6)	4) (−95) ÷ (− 5)
5) 32 × (− 4)	6) (− 99) ÷ (− 11)
7) (− 12) × (− 4)	8) (− 123) ÷ 1
9) (− 4) × (− 3) × 5	10) (− 0) ÷ 15

Score: ....

Answer Key

1) 40	2) 8
3) 120	4) 19
5) −128	6) 9
7) 48	8) −123
9) 60	10) 0

TASC Math Prep

Name: ....

Arrange, Order, and Comparing Integers

- ✓ When we use a number line, numbers are increased when you go to the right.
- ✓ To compare numbers, you can use number line! As you go from left to right on the number line, you will find a greater number!
- ✓ Order integers from smallest to greatest.

EXAMPLE:

Order integers from to greatest.

$$(-11, -13, 7, -2, 12)$$

To compare numbers, you can use number line! When you see from left to right on the number line, you find a greater number!

$-13 < -11 < -2 < 7 < 12$

PRACTICES:

Order each set of integers from least to greatest.	Order each set of integers from greatest to least
1) $2, 6, -15, -11, 1$	2) $1, 17, 6, 8, 65, 2$
3) $9, -8, 3, -2, 11$	4) $-12, 6, -7, 2, -11$
5) $36, -12, 5, 1, -2$	6) $-14, 17, 7, 37, 9$
7) $31, 18, 0, -54, 9, -5$	8) $-54, 0, 14, 19, 15$
9) $-15, -25, -37, 7, 0, 9$	10) $12, 7, -1, -11, 9, -3$

Score: ....

Answer Key

1) $-15, -11, 1, 2, 6$	2) $65, 17, 8, 6, 2, 1$
3) $-8, -2, 3, 9, 11$	4) $6, 2, -7, -11, -12$
5) $-12, -2, 1, 5, 36$	6) $37, 17, 9, 7, -14$
7) $-54, -5, 0, 9, 18, 31$	8) $19, 15, 14, 0, -54$
9) $-37, -25, -15, 0, 7, 9$	10) $12, 9, 7, -1, -3, -11$

Compare Integer

- If you want to compare numbers, you can use a number line! As you move from left to right on the number line, you will find a greater number!

EXAMPLE:

-5 ____ -1

When we compare two integers, we use the symbols < and >.

$-5 < -1$ means that -5 is less than -1

PRACTICES:

Compare. Use >, =, <

1) 4 ____ 3

2) -22 ___ -11

3) 0 ____ -31

4) -41 ____ -12

5) -64 ____ 64

6) -142 ____ -148

7) 68 ____ 100

8) $(-15) \times 6$ ____ $5 \times (-18)$

9) 16 ____ $-(-16)$

10) 405 ____ -405

Score: ....

Answer Key

1) >	2) <
3) >	4) <
5) <	6) >
7) <	8) =
9) =	10) >

Order of Operations

When you find more than one math operation, use PEMDAS:
- ✓ Parentheses
- ✓ Exponents
- ✓ Multiplication and Division (from left to right)
- ✓ Addition and Subtraction (from left to right)

EXAMPLE:

Solve. $(11 \times 5) - (12 - 25) =$

First you have to simplify inside parentheses: $(11 \times 5) - (12 - 25) = (55) - (-13) =$

Then: $55 + 13 = 68$

PRACTICES:

Evaluate each expression.

1) $24 - (8 \times 6)$

2) $5 \times 6 - (\frac{15}{11 - (-4)})$

3) $12 - (6 \times (-3))$

4) $(6 \times 7) + (-7)$

5) $(\frac{(-1)+4}{(-1)+(-2)}) \times (-9)$

6) $\frac{30}{2(9-(-1))-10}$

7) $58 - (6 \times 9)$

8) $13 + (4 \times 2)$

9) $((-3) + 15) \div (-3)$

10) $[(-8 \div 2) \div (2 - 4))$

TASC Math Prep

Score: ....

Answer Key

1) −24	2) 29
3) 30	4) 35
5) 9	6) 3
7) 4	8) 21
9) −4	10) 2

Integers and Absolute Value

- To find a definite value of a number, simply find its distance from 0 on number line! For example, the distance of 13 and −13 from zero on number line is 13!

EXAMPLE:

Solve. $\frac{|-18|}{9} \times |5-8| =$

First find $|-18|$, → the definite value of −18 is 18, then: $|-18| = 18$

$\frac{18}{9} \times |5-8| =$

Next, we solve $|5-8|$, → $|5-8| = |-3|$, the definite value of −3 is 3. $|-3| = 3$

Then: $\frac{18}{9} \times 3 = 2 \times 3 = 6$

PRACTICES:

Write absolute value of each number.

1) 62	2) −32
3) −11	4) 5

Evaluate.

5) $\|-12\| - \|3\| + 2$	6) $19 + \|-5 - 14\| - \|2\|$
7) $\|-11\| + \|-9\|$	8) $\|91\| - \|-18\| - 18$
9) $\|-10 + 4\| \times \frac{\|-7 \times 5\|}{7}$	10) $\frac{\|-16 \times 3\|}{2} \times \|-12\|$

Score: ....

Answer Key

1) 62	2) 32
3) 11	4) 5
5) 11	6) 36
7) 20	8) 55
9) 30	10) 288

Chapter 2 : Fractions and Decimals

Topics that you'll learn in this chapter:

- Simplifying Fractions
- Adding and Subtracting Fractions, Mixed Numbers and Decimals
- Multiplying and Dividing Fractions, Mixed Numbers and Decimals
- Comparing and Rounding Decimals
- Converting Between Fractions, Decimals and Mixed Numbers
- Factoring Numbers, Greatest Common Factor, and Least Common Multiple
- Divisibility Rules

"A Man is like a fraction whose numerator is what he is and whose denominator is what he thinks of himself. The larger the denominator, the smaller the fraction." –Tolstoy

Simplifying Fractions

- ✓ Regularly divide both the top and bottom of the fraction by 2, 3, 5, 7, ... etc.
- ✓ Continue until you can't go any further.

EXAMPLE:

Simplify $\frac{12}{20}$.

To simplify $\frac{12}{20}$, you have to find a number that both 12 and 20 are divisible by. Both are divisible by 4. Then: $\frac{12}{20} = \frac{12 \div 4}{20 \div 4} = \frac{3}{5}$

PRACTICES:

Simplify the fractions.

1) $\frac{44}{64}$

2) $\frac{12}{26}$

3) $\frac{15}{25}$

4) $\frac{30}{45}$

5) $\frac{18}{27}$

6) $1\frac{62}{124}$

7) $4\frac{12}{66}$

8) $1\frac{55}{70}$

9) $\frac{54}{60}$

10) $7\frac{68}{136}$

Score: ....

Answer Key

1) $\frac{11}{16}$	2) $\frac{6}{13}$
3) $\frac{3}{5}$	4) $\frac{2}{3}$
5) $\frac{2}{3}$	6) $1\frac{1}{2}$
7) $4\frac{2}{11}$	8) $1\frac{11}{14}$
9) $\frac{9}{10}$	10) $7\frac{1}{2}$

Factoring Numbers

- ✓ To break the numbers into their prime factors is called factoring.
- ✓ First few prime numbers are 2, 3, 5, 7, 11, 13, 17, 19

EXAMPLE:

List all positive factors of 12.

Write the upside-down division:

The second column is the answer.

Then: $12 = 2 \times 2 \times 3$ or $12 = 2^2 \times 3$

12	2
6	2
3	3
1	

PRACTICES:

List all positive factors of each number.	List the prime factorization for each number.
1) 90	2) 40
3) 49	4) 105
5) 50	6) 42
7) 34	8) 78
9) 96	10) 165

Answer Key

1) 1, 2, 3, 5, 6, 9, 10, 15, 18, 30, 45, 90	2) $2 \times 2 \times 2 \times 5$
3) 1, 7, 49	4) $3 \times 5 \times 7$
5) 1, 2, 5, 10, 25, 50	6) $2 \times 3 \times 7$
7) 1, 2, 17, 34	8) $2 \times 3 \times 13$
9) 1, 2, 3, 4, 6, 8, 12, 16, 24, 32, 48, 96	10) $3 \times 5 \times 11$

Greatest Common Factor (GCF)

- ✓ List the prime factors of each number.
- ✓ Then multiply common prime factors.
- ✓ If there are no common prime factors, then our GCF is 1.

EXAMPLE:

Find the GCF for **10** and **15**.

The factors of 10 are: {1, 2, 5, 10}

The factors of 15 are: {1, 3, 5, 15}

There is 5 in common,

Then the greatest common factor is: 5

PRACTICES:

Find the GCF for each number pair.

1) 12, 25	2) 72, 84
3) 24, 36	4) 30, 45
5) 9, 36	6) 63, 42
7) 27, 12	8) 125, 50
9) 54, 39	10) 36, 52

Answer Key

1) 1	2) 12
3) 12	4) 15
5) 9	6) 21
7) 3	8) 25
9) 3	10) 4

Least Common Multiple (LCM)

- ✓ The smallest multiple that 2 or more numbers have in common is called least common multiple of that number. How to find LCM:
- ✓ First find the list of the prime factors of each number.
- ✓ Then multiply the common prime factors and uncommon prime factors of the numbers (each common prime factor is used only for once)

EXAMPLE:

Find the LCM for **18** and **12**.

Prime Numbers: 12 = 2 × 2 × 3   18 = 2 × 3 × 3

Common factors: 2 and 3
Not Common: 2 and 3

2 × 2 × 3 × 3 = 36

PRACTICES:

Find the LCM for each number pair.

1) 12, 9	2) 40, 20
3) 15, 30	4) 84, 60
5) 60, 40	6) 52, 78
7) 14, 28	8) 14, 7, 42
9) 24, 32	10) 72, 66, 24

Score: ....

Answer Key

1) 36	2) 40
3) 30	4) 420
5) 120	6) 156
7) 28	8) 42
9) 96	10) 792

TASC Math Prep

Name: ....

Divisibility Rules

If a number can be divided by other numbers, it is referred as divisibility. The number is divisible:
- ✓ by 2 if the number is found even.
- ✓ by 3 if the sum of the digits is found to be divisible by 3.
- ✓ by 9 if the sum of the digits is found to be divisible by 9.
- ✓ by 4, if the last 2 digits of a number are found to be divisible by 4.
- ✓ by 6, if it is found to be divisible by 2 and 3.
- ✓ by 8, if it is found to be divisible by 2 and 4.
- ✓ by 5 if the last digit is found 0 or 5.
- ✓ by 10 if the last digit is 0.

EXAMPLE:

What is the factor of 240?

2, because the number is even

3, because $(2 + 4 + 0 = 6, 6 \div 3 = 2)$

4, because $(40 \div 4 = 10)$

5, because the last digit is 0

8, because of 2 and 4

10, because the last digit is 0

Then 240 is divisible by 2, 3, 4, 5, 8, 10

PRACTICES:

Use the divisibility rules to find the factors of each number.

1) 12	2 3 4 5 6 7 8 9 10	2) 24	2 3 4 5 6 7 8 9 10
3) 36	2 3 4 5 6 7 8 9 10	4) 18	2 3 4 5 6 7 8 9 10
5) 30	2 3 4 5 6 7 8 9 10	6) 54	2 3 4 5 6 7 8 9 10
7) 90	2 3 4 5 6 7 8 9 10	8) 80	2 3 4 5 6 7 8 9 10
9) 72	2 3 4 5 6 7 8 9 10	10) 84	2 3 4 5 6 7 8 9 10

www.mathnotion.com

Answer Key

1) 12	2 3 4 5 6 7 8 9 10	2) 24	2 3 4 5 6 7 8 9 10
3) 36	2 3 4 5 6 7 8 9 10	4) 18	2 3 4 5 6 7 8 9 10
5) 30	2 3 4 5 6 7 8 9 10	6) 54	2 3 4 5 6 7 8 9 10
7) 90	2 3 4 5 6 7 8 9 10	8) 80	2 3 4 5 6 7 8 9 10
9) 72	2 3 4 5 6 7 8 9 10	10) 84	2 3 4 5 6 7 8 9 10

Adding and Subtracting Fractions

- ✓ Find equivalent fractions with the equivalent divisor before you can add or subtract fractions with totally different divisors.
- ✓ Adding and Subtracting with the equivalent divisors:

  $$\frac{a}{b} + \frac{c}{b} = \frac{a+c}{b}, \frac{a}{b} - \frac{c}{b} = \frac{a-c}{b}$$

- ✓ Adding and Subtracting fractions with different divisors:

  $$\frac{a}{b} + \frac{c}{d} = \frac{ad+cb}{bd}, \frac{a}{b} - \frac{c}{d} = \frac{ad-cb}{bd}$$

EXAMPLE:

Subtract fractions. $\frac{2}{3} - \frac{1}{2} = ?$

For "unlike" fractions, find equivalent fractions with the same divisors before you can add or subtract fractions with different divisors. Use this formula: $\frac{a}{b} - \frac{c}{d} = \frac{ad-cb}{bd}$

$$\frac{2}{3} - \frac{1}{2} = \frac{(2)(2) - (1)(3)}{3 \times 2} = \frac{4-3}{6} = \frac{1}{6}$$

PRACTICES:

Add fractions.	Subtract fractions.
1) $\frac{1}{4} + \frac{2}{3}$	2) $\frac{1}{2} - \frac{1}{5}$
3) $\frac{1}{3} + \frac{1}{2}$	4) $\frac{1}{7} - \frac{1}{9}$
5) $\frac{1}{4} + \frac{5}{7}$	6) $\frac{3}{5} - \frac{1}{15}$
7) $\frac{6}{7} + \frac{3}{21}$	8) $\frac{1}{3} - \frac{1}{4}$
9) $\frac{5}{13} + \frac{1}{2}$	10) $\frac{6}{5} - \frac{5}{6}$

Score: ....

Answer Key

1) $\frac{11}{12}$	2) $\frac{3}{10}$
3) $\frac{5}{6}$	4) $\frac{2}{63}$
5) $\frac{27}{28}$	6) $\frac{8}{15}$
7) 1	8) $\frac{1}{12}$
9) $\frac{23}{26}$	10) $\frac{11}{30}$

Multiplying and Dividing Fractions

- ✓ How to multiply fractions: multiply the top numbers and multiply the bottom numbers.
- ✓ How to change fractions: Keep, Change, Flip
- ✓ Keep the first fraction then change division sign into multiplication sign and flip the numerator and denominator of the second fraction. Then, solve it

EXAMPLE:

Multiplying fractions. $\frac{5}{6} \times \frac{3}{4} =$

Multiply the upper numbers and multiply the lower numbers.

$\frac{5}{6} \times \frac{3}{4} = \frac{5 \times 3}{6 \times 4} = \frac{15}{24}$, simplify: $\frac{15}{24} = \frac{15 \div 3}{24 \div 3} = \frac{5}{8}$

Dividing fractions. $\frac{1}{4} \div \frac{2}{3} =$

Keep the first fraction then change division sign into multiplication sign and flip the numerator and denominator of the second fraction. Then: $\frac{1}{4} \times \frac{3}{2} = \frac{1 \times 3}{4 \times 2} = \frac{3}{8}$

PRACTICES:

Multiplying fractions. Then simplify.	Dividing fractions.
1) $\frac{3}{5} \times \frac{5}{9}$	2) $\frac{4}{9} \div 4$
3) $\frac{5}{21} \times \frac{7}{10}$	4) $\frac{32}{25} \div \frac{8}{5}$
5) $\frac{3}{29} \times \frac{29}{3}$	6) $\frac{2}{7} \div \frac{8}{35}$
7) $\frac{8}{11} \times 11$	8) $\frac{12}{25} \div \frac{3}{5}$
9) $\frac{7}{9} \times \frac{12}{28}$	10) $7 \div \frac{2}{3}$

Score: ....

Answer Key

1) $\frac{1}{3}$	2) $\frac{1}{9}$
3) $\frac{1}{6}$	4) $\frac{4}{5}$
5) 1	6) $1\frac{1}{4}$
7) 8	8) $\frac{4}{5}$
9) $\frac{1}{3}$	10) $10\frac{1}{2}$

Adding Mixed Numbers

Use these steps for both adding and subtracting mixed numbers.
- ✓ Add whole number of the mixed numbers.
- ✓ Add the fractions of every mixed number.
- ✓ Find the Least Common Divisor (LCD) if needed.
- ✓ Add whole numbers and fractions.
- ✓ Write your answer in simplest form.

EXAMPLE:

$1\frac{3}{4} + 2\frac{3}{8} = ?$

Rewriting our equation with parts separated, $1 + \frac{3}{4} + 2 + \frac{3}{8}$, Solving the whole number parts $1 + 2 = 3$, Solving the fraction parts $\frac{3}{4} + \frac{3}{8}$, and rewrite to solve with the equivalent fractions.

$\frac{6}{8} + \frac{3}{8} = \frac{9}{8} = 1\frac{1}{8}$, then combining the whole and fraction parts $3 + 1 + \frac{1}{8} = 4\frac{1}{8}$

PRACTICES:

Add.

1) $1\frac{1}{7} + 2\frac{1}{3}$	2) $1\frac{1}{2} + 3\frac{2}{3}$
3) $1\frac{2}{5} + 2\frac{1}{10}$	4) $7 + 2\frac{1}{2}$
5) $4\frac{1}{3} + 2\frac{2}{3}$	6) $2\frac{2}{3} + 1\frac{1}{4}$
7) $2\frac{3}{4} + 3\frac{1}{8}$	8) $9 + 1\frac{1}{9}$
9) $4\frac{5}{12} + 2\frac{3}{4}$	10) $3\frac{1}{7} + 2\frac{3}{14}$

Score: ....

Answer Key

1) $3\frac{10}{21}$	2) $5\frac{1}{6}$
3) $3\frac{1}{2}$	4) $9\frac{1}{2}$
5) 7	6) $3\frac{11}{12}$
7) $5\frac{7}{8}$	8) $10\frac{1}{9}$
9) $7\frac{1}{6}$	10) $5\frac{5}{14}$

Subtracting Mixed Numbers

Use these steps for both adding and subtracting mixed numbers.
- ✓ From whole number of the first mixed number, subtract the whole number of second mixed number.
- ✓ From first fraction subtract the second.
- ✓ Find the Least Common Divisor (LCD) if needed.
- ✓ Add the result of whole numbers and fractions.
- ✓ Write your answer in simplest terms.

EXAMPLE:

$5\frac{2}{3} - 2\frac{1}{4} = ?$

Rewriting our equation with parts separated, $5 + \frac{2}{3} - 2 - \frac{1}{4}$

Solving the whole number parts $5 - 2 = 3$, Solving the fraction parts, $\frac{2}{3} - \frac{1}{4} = \frac{8-3}{12} = \frac{5}{12}$

Joining the whole and fraction parts, $3 + \frac{5}{12} = 3\frac{5}{12}$

PRACTICES:

Subtract.

1) $5\frac{2}{7} - 2\frac{1}{14}$

2) $4\frac{2}{5} - \frac{2}{3}$

3) $3\frac{3}{7} - 1\frac{1}{14}$

4) $7\frac{2}{5} - 5\frac{1}{3}$

5) $4\frac{1}{2} - 2\frac{4}{8}$

6) $11\frac{5}{12} - 8\frac{3}{4}$

7) $7\frac{5}{12} - 5\frac{7}{12}$

8) $5\frac{2}{9} - 2\frac{1}{18}$

9) $3\frac{2}{5} - 2\frac{1}{5}$

10) $3\frac{4}{9} - 1\frac{2}{9}$

TASC Math Prep

Score: ....

Answer Key

1) $3\frac{3}{14}$	2) $3\frac{11}{15}$
3) $2\frac{5}{14}$	4) $2\frac{1}{15}$
5) 2	6) $2\frac{2}{3}$
7) $1\frac{5}{6}$	8) $3\frac{1}{6}$
9) $1\frac{1}{5}$	10) $2\frac{2}{9}$

Multiplying Mixed Numbers

- ✓ Convert the mixed numbers into improper fractions. (Improper fraction is a fraction in which the numerator is greater than denominator)
- ✓ Multiply fractions and write in simplest form if needed.

$$a\frac{c}{b} = a + \frac{c}{b} = \frac{ab+c}{b}$$

EXAMPLE:

Multiply mixed numbers. $4\frac{3}{5} \times 2\frac{1}{3} = ?$

Changing mixed numbers to fractions, $\frac{23}{5} \times \frac{7}{3}$, Applying the fractions formula for multiplication, $\frac{23 \times 7}{5 \times 3} = \frac{161}{15} = 10\frac{11}{15}$

PRACTICES:

Find each product.

1) $2\frac{1}{3} \times \frac{1}{2}$	2) $1\frac{2}{5} \times \frac{2}{3}$
3) $2\frac{4}{3} \times 2\frac{2}{6}$	4) $2\frac{1}{2} \times 1\frac{2}{4}$
5) $3\frac{1}{2} \times 1\frac{2}{3}$	6) $1\frac{1}{7} \times 1\frac{3}{4}$
7) $1\frac{1}{4} \times 2\frac{6}{5}$	8) $3\frac{1}{2} \times 4\frac{2}{5}$
9) $1\frac{2}{5} \times 2\frac{1}{3}$	10) $5\frac{7}{12} \times 2\frac{4}{9}$

Score: ....

Answer Key

1) $1\frac{1}{6}$	2) $\frac{14}{15}$
3) $7\frac{7}{9}$	4) $3\frac{3}{4}$
5) $5\frac{5}{6}$	6) 2
7) 4	8) $15\frac{2}{5}$
9) $3\frac{4}{15}$	10) $13\frac{35}{54}$

Dividing Mixed Numbers

- ✓ Change the mixed numbers into improper fractions.
- ✓ Divide fractions and write in simplest form if needed.

$$a\frac{c}{b} = a + \frac{c}{b} = \frac{ab+c}{b}$$

EXAMPLE:

Find the quotient. $2\frac{1}{2} \div 1\frac{1}{5} = ?$

Changing mixed numbers to fractions, $\frac{5}{2} \div \frac{6}{5}$, Applying the fractions formula for multiplication, $\frac{5}{2} \times \frac{5}{6} = \frac{5 \times 5}{2 \times 6} = \frac{25}{12} = 2\frac{1}{12}$

PRACTICES:

Find each quotient.

1) $2\frac{3}{5} \div 1\frac{3}{8}$

2) $\frac{3}{2} \div 2\frac{3}{4}$

3) $1\frac{4}{7} \div 2\frac{2}{3}$

4) $1\frac{2}{3} \div 2\frac{1}{3}$

5) $0 \div 4\frac{2}{5}$

6) $2\frac{2}{5} \div 1\frac{1}{2}$

7) $1\frac{2}{3} \div 2\frac{1}{5}$

8) $3\frac{2}{7} \div 4\frac{3}{5}$

9) $1\frac{1}{4} \div 2\frac{4}{5}$

10) $2 \div 3\frac{1}{3}$

Score: ....

Answer Key

1) $1\frac{49}{55}$	2) $\frac{6}{11}$
3) $\frac{33}{56}$	4) $\frac{5}{7}$
5) 0	6) $1\frac{3}{5}$
7) $\frac{25}{33}$	8) $\frac{5}{7}$
9) $\frac{25}{56}$	10) $\frac{3}{5}$

TASC Math Prep

Name: ....

Comparing Decimals

Decimal is a fraction written in a unique form. For example, instead of writing $\frac{1}{2}$ you can write as 0.5.

For comparison of decimals:

✓ Compare every digit of two decimals in the same place value.
✓ Start from left. Compare ones, tens, hundreds, tenths, hundredths, etc.
✓ To compare numbers, use these symbols:
- Equal to =, Less than <, Greater than >
- Greater than or equal ≥, Less than or equal ≤

EXAMPLE:

Compare 0.20 and 0.02.

0.20 is greater than 0.02, because the tenth place of 0.20 is 2, but the tenth place of 0.02 is zero. Then: $0.20 > 0.02$

PRACTICES:

Write the correct comparison symbol (>, < or =).

1) 0.025 ____ 0.25	2) 0.9 ____ 0.888
3) 4.510 ____ 4.150	4) 10.01 ____ 10.10
5) 0.987 ____ 0.991	6) 18.004 ____ 18.040
7) 0.020 ____ 0.20	8) 0.071 ____ 0.700
9) 0.08 ____ 0.009	10) 0.690 ____ 0.609

Score: ....

Answer Key

1) <	2) >
3) >	4) <
5) <	6) <
7) <	8) <
9) >	10) >

Name: ....

Rounding Decimals

- ✓ To round a decimal, you must find the place value you'll round to.
- ✓ Then, find the digit to the right of the place value you're rounding to.
  - If it is 5 or greater, add 1 to the place value you're rounding to and remove all digits on its right side.
  - If the digit to the right of the place value is smaller than 5, keep the place value and remove all digits on the right.

EXAMPLE:

Round 2.1837 to the thousandth-place value.

First have a look at the next place value to the right, (tens thousandths). It's 7 and it's found to be greater than 5. So, add 1 to the digit in the thousandth place.

Thousandth place is 3. → 3 + 1 = 4, then, the answer is 2.184

PRACTICES:

Round each decimal number to the nearest place indicated.

1) 6.0̲8	2) 12.2̲67
3) 9.3̲01	4) 10.07̲1
5) 55̲.89	6) 59̲.15
7) 32̲9.018	8) 92.41̲0
9) 1.49̲9	10) 25̲.621

www.mathnotion.com

Score: ....

Answer Key

1) 6.1	2) 12.3
3) 9.3	4) 10.07
5) 56	6) 59
7) 330	8) 92.4
9) 1.5	10) 26

Adding and Subtracting Decimals

- ✓ Arrange the numbers in line.
- ✓ Add zeros to have same number of digits for both the numbers.
- ✓ Add or subtract by using column subtraction or addition.

EXAMPLE:

Add. $2.5 + 1.24 =$

First line up the numbers: $\begin{array}{r} 2.5 \\ +1.24 \\ \hline \end{array}$ → Add zeros to have same number of digits for both numbers. $\begin{array}{r} 2.50 \\ +1.24 \\ \hline \end{array}$, Start with the hundredths place. $0 + 4 = 4$, $\begin{array}{r} 2.50 \\ +1.24 \\ \hline 4 \end{array}$, Continue with tenths place. $5 + 2 = 7$, $\begin{array}{r} 2.50 \\ +1.24 \\ \hline .74 \end{array}$. Add the ones place. $2 + 1 = 3$, $\begin{array}{r} 2.50 \\ +1.24 \\ \hline 3.74 \end{array}$

Subtract decimals. $4.67 - 2.15 = \begin{array}{r} 4.67 \\ -2.15 \\ \hline \end{array}$

Start with the hundredths place. $7 - 5 = 2$, $\begin{array}{r} 4.67 \\ -2.15 \\ \hline 2 \end{array}$, continue with tenths place. $6 - 1 = 5$ $\begin{array}{r} 4.67 \\ -2.15 \\ \hline .52 \end{array}$, subtract the ones place. $4 - 2 = 2$, $\begin{array}{r} 4.67 \\ -2.15 \\ \hline 2.52 \end{array}$.

PRACTICES:

Add and subtract decimals.

1)	87.15 − 32.35	2)	90.43 + 44.09
3)	58.56 + 12.10	4)	65.23 − 56.48
5)	98.125 + 58.54	6)	162.05 − 83.65

Solve.

7) ___ + 5.0 = 9.08

8) 7.06 + ___ = 24.6

9) 21.9 − ___ = 6.9

10) 32.12 − ___ = 12.07

Score: ....

Answer Key

1) 54.8	2) 134.52
3) 70.66	4) 8.75
5) 156.665	6) 78.4
7) 4.08	8) 17.54
9) 15	10) 20.05

TASC Math Prep

Name: ....

Multiplying Decimals

- ✓ Arrange and multiply the numbers as you do with whole numbers.
- ✓ Then count the total number of decimal places in every factor.
- ✓ Place the decimal point in the product.

EXAMPLE:

Find the product. $0.50 \times 0.20 =$

Arrange and multiply the numbers as you do with whole numbers. Line up the numbers:

$\frac{50}{\times 20}$, Start with the ones place → $0 \times 50 = 0$, $\frac{50}{\times 20} \atop 0$, Continue with other digits →

$2 \times 50 = 100$, $\frac{50}{\times 20} \atop 1,000$, Count the total number of decimal places in both of the factors

(4). Then Place the decimal point in the product.

Then: $\frac{0.50}{\times 0.20} \atop 0.1000$ → $0.50 \times 0.20 = 0.1$

PRACTICES:

Find each product.

1) $\quad1.5 \atop \underline{\times 0.16}$	2) $\quad5.3 \atop \underline{\times 1.9}$
3) $\quad0.06 \atop \underline{\times 2.5}$	4) $\quad3.19 \atop \underline{\times 21.5}$
5) $\quad9.3 \atop \underline{\times 11.5}$	6) $\quad3.01 \atop \underline{\times 2.1}$
7) $\quad5.0 \atop \underline{\times 1.4}$	8) $\quad23.8 \atop \underline{\times 10}$
9) $\quad21.5 \atop \underline{\times 0.001}$	10) $\quad8.21 \atop \underline{\times 3.1}$

www.mathnotion.com

Score: ....

Answer Key

1) 0.24	2) 10.07
3) 0.15	4) 68.585
5) 106.95	6) 6.321
7) 7	8) 238
9) 0.0215	10) 25.451

Name: ....

Dividing Decimals

- ✓ If the divisor is not a whole number, transfer decimal point to right to make it a whole number. Do the same step for dividend.
- ✓ Divide same to whole numbers.

EXAMPLE:

Find the quotient. $1.20 \div 0.2 =$

The divisor is not a whole number. Multiply it by 10 to get 2. Do the same step for the dividend to get 12. Now, divide: $12 \div 2 = 6$. The answer is 6.

PRACTICES:

Find each quotient.

1) $25.7 \div 0.5$	2) $67.2 \div 4$
3) $61.75 \div 1.9$	4) $18.0 \div 1.2$
5) $12.4 \div 10$	6) $2.2 \div 100$
7) $1.88 \div 100$	8) $55.1 \div 100$
9) $0.1 \div 100$	10) $0.25 \div 10$

Score: ....

Answer Key

1) 51.4	2) 16.8
3) 32.5	4) 15
5) 1.24	6) 0.022
7) 0.0188	8) 0.551
9) 0.001	10) 0.025

Converting Between Fractions, Decimals and Mixed

How to convert fraction into Decimal:
- ✓ Divide the numerator by denominator.

How to convert decimal into Fraction:
- ✓ Write decimal over 1.
- ✓ Multiply both numerator value and denominator value by 10 for each digit on the right side of the decimal point.
- ✓ Make it to simplest form.

EXAMPLE:

What is long division of $\frac{5}{8}=$?

In that case we put extra zeros and did $\frac{5.000}{8}$ to get 0.625

PRACTICES:

Convert fractions to decimals.	Convert decimal into fraction or mixed numbers
1) $\frac{4}{10}$	2) 3.6
3) $\frac{3}{8}$	4) 0.07
5) $\frac{4}{12}$	6) 0.15
7) $\frac{5}{16}$	8) 2.7
9) $\frac{60}{100}$	10) 2.5

Score: ....

Answer Key

1) 0.4	2) $3\frac{3}{5}$
3) 0.375	4) $\frac{7}{100}$
5) 0.333	6) $\frac{3}{20}$
7) 0.3125	8) $2\frac{7}{10}$
9) 0.6	10) $2\frac{1}{2}$

Chapter 3 : Proportion, Ratio, Percent

Topics that you'll learn in this chapter:

- Writing and Simplifying Ratios
- Create a Proportion
- Similar Figures
- Simple Interest
- Ratio and Rates Word Problems
- Percentage Calculations
- Converting Between Percent, Fractions, and Decimals
- Percent Problems
- Markup, Discount, and Tax

"Do not worry about your difficulties in mathematics. I can assure you mine are still greater." – *Albert Einstein*

Name: ....

Writing Ratios

✓ A ratio is a comparison of two numbers, and it can be written as a division.

EXAMPLE:

$3 : 5 = ?$

Both numbers 3 and 5 are divisible by 8, $\Rightarrow 3 \div 8 = \frac{3}{8}, 5 \div 8 = \frac{5}{8}$,

Then: $3 : 5 = \frac{3}{8}$ and $\frac{5}{8}$.

PRACTICES:

Express each ratio as a rate and unite rate.	Express each ratio as a fraction in the simplest form
1) 80 dollars for 4 chairs.	2) 13 cups to 39 cups.
3) 125 miles on 25 gallons of gas.	4) 17 cakes out of 51 cakes
5) 147 miles on 7 hours.	6) 35 red desks out of 125 desks
7) 12 inches of snow in 24 hours.	8) 8 story books out of 32 books
9) 14 dimes to 112 dimes.	10) 12 gallons to 20 gallons

Score: ....

Answer Key

1)	$\frac{80 \text{ dollars}}{4 \text{ books}}$, 20.00 dollars per chair	2)	$\frac{1}{3}$
3)	$\frac{125 \text{ miles}}{25 \text{ gallons}}$, 5 miles per gallon	4)	$\frac{1}{3}$
5)	$\frac{147 \text{ miles}}{7 \text{ hours}}$, 21 miles per hour	6)	$\frac{7}{25}$
7)	$\frac{12" \text{ of snow}}{24 \text{ hours}}$, 0.5 inches of snow per hour	8)	$\frac{1}{4}$
9)	$\frac{14 \text{ dimes}}{112 \text{ dimes}}$, $\frac{1}{8}$ per dime	10)	$\frac{3}{5}$

Simplifying Ratios

- ✓ Ratios are used to compare two numbers.
- ✓ Ratios can be written as a fraction, using colon or the word "to".
- ✓ You can calculate identical ratios by multiplying or dividing both sides of the ratio by the same number.

EXAMPLE:

Simplify. $8:4 =$

Both numbers 8 and 4 are divisible by 4, $\Rightarrow 8 \div 4 = 2, 4 \div 4 = 1,$

Then: $8:4 = 2:1$

PRACTICES:

Reduce each ratio.

1) 49: 14	2) 22: 55
3) 35: 25	4) 18: 99
5) 16: 36	6) 64: 72
7) 4: 60	8) 70: 40
9) 8: 64	10) 16: 24

Score: ....

Answer Key

1) 7: 2	2) 2: 5
3) 7: 5	4) 2: 11
5) 4: 9	6) 8: 9
7) 1: 15	8) 7: 4
9) 1: 8	10) 2: 3

Create a Proportion

- A proportion carries two equal fractions! A proportion means equality of two fractions.
- If you want to create a proportion, simply find (or create) two equal fractions.

EXAMPLE:

Explain if these ratios form a proportion. $\frac{3}{5}$ and $\frac{24}{45}$

Use cross multiplication: $\frac{3}{5} = \frac{24}{45} \rightarrow 3 \times 45 = 5 \times 24 \rightarrow 135 = 120$, which is not correct. Thus, this pair of ratios doesn't form a proportion.

PRACTICES:

Create proportion from the given set of numbers.

1) 3, 2, 9, 6	2) 4, 18, 12, 6
3) 5, 11, 25, 55	4) 24, 7, 21, 8
5) 49, 7, 12, 84	6) 15, 12, 30, 24
7) 20, 10, 200, 1	8) 9, 27, 81, 3
9) 4, 2, 16, 32	10) 9, 6, 27, 18

Score: ....

Answer Key

1) 2: 6 = 3: 9	2) 4: 12 = 6: 18
3) 5: 25 = 11: 55	4) 8: 24 = 7: 21
5) 7: 49 = 12: 84	6) 12: 24 = 15: 30
7) 1: 10 = 20: 200	8) 3: 27 = 9: 81
9) 2: 16 = 4: 32	10) 6: 18 = 9: 27

Similar Figures

- Two or more figures are equivalent if their corresponding angles are equal, and the corresponding sides are in proportion.

EXAMPLE:

4–5–6 triangle is like an 8–10–12 triangle.

PRACTICES:

Each pair of figures is similar. Find the missing side.

1)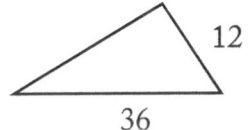

2)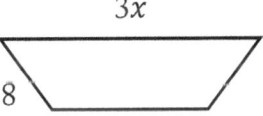

3)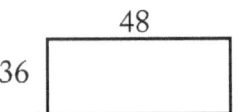

4)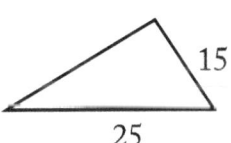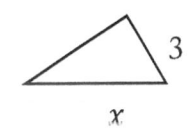

5)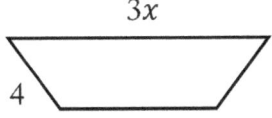

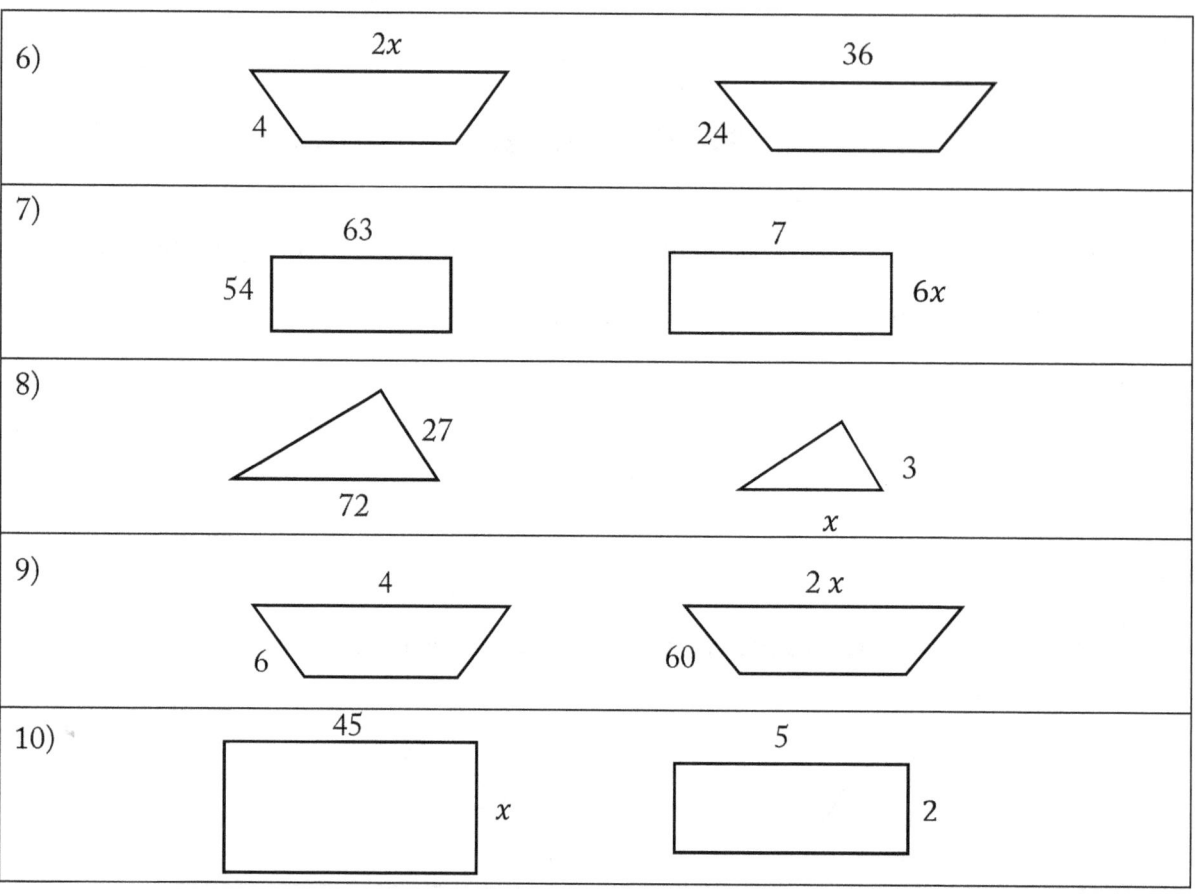

Score: ....

Answer Key

1) 6	2) 4
3) 3	4) 5
5) 2	6) 3
7) 1	8) 8
9) 20	10) 18

Name: ....

Ratio and Rates Word Problems

✓ To solve a rate word problem or a ratio, create a proportion and then use cross multiplication method.

EXAMPLE:

A tree **32 feet** tall has a shadow **12 feet** long. Jack is **6 feet** tall. How long is Jack's shadow?

To solve for the missing number, write in a proportion.

$\frac{32}{12} = \frac{6}{x} \rightarrow 32x = 6 \times 12 = 72$

$32x = 72 \rightarrow x = \frac{72}{32} = 2.25$

PRACTICES:

Solve.

1) In a party, 8 soft drinks are required for every 35 guests. If there are 560 guests, how many soft drinks is required?

2) You can buy 6 cans of green beans at a supermarket for $3.50. How much does it cost to buy 42 cans of green beans?

3) The price of 5 bananas at the first Market is $1.05. The price of 7 of the same bananas at second Market is $1.07. Which place is the better buy?

4) In Peter's class, 21 of the students are tall and 9 are short. In Elise's class 56 students are tall and 24 students are short. Which class has a higher ratio of tall to short students?

5) The bakers at a Bakery can make 110 bagels in 4 hours. How many bagels can they bake in 6 hours? What is that rate per hour?

6) A certain sweet recipe calls for 3 kg of sugar for every 6 kg of flour. If 63 kg of this sweet must be prepared, how much sugar is required?

7) In a mixture of 45 liters, the ratio of sugar solution to salt solution is 1:2. What is the amount of sugar solution to be added if the ratio must be 2:1?

8) In a bag of red and green sweets, the ratio of red sweets to green sweets is 3:4. If the bag contains 120 green sweets, how many red sweets are there?

9) If the ratio of chocolates to ice-cream cones in a box is 5:8 and the number of chocolates is 30, find the number of ice-cream cones.

10) In a group, the ratio of doctors to lawyers is 5:4. If the total number of people in the group is 72, what is the number of lawyers in the group?

Score: ....

Answer Key

1) 128	2) $24.5
3) The price at the second Market is a better buy.	4) The ratio for both classes equal 7 to 3.
5) 165, the rate is 27.5 per hour.	6) 21 kg (3+6=9, $\frac{63}{9}$ = 7. Therefore, 3:6=21:42)
7) 45	8) 90
9) 48	10) 32

Name: ....

Percentage Calculations

- ✓ Percent is called the ratio of a number and 100. It always possesses the same denominator, 100. The symbol used for percent is %.
- ✓ Percent is another method to write decimals or fractions. For example:
  $40\% = 0.40 = \frac{40}{100} = \frac{2}{5}$
- ✓ Use the given formula to find part, whole, or percent:
  $\text{part} = \frac{\text{percent}}{100} \times \text{whole}$

EXAMPLE:

What is 10% of 45?

Use this formula: $\text{part} = \frac{\text{percent}}{100} \times \text{whole}$

$\text{part} = \frac{10}{100} \times 45 \rightarrow \text{part} = \frac{1}{10} \times 45 \rightarrow \text{part} = \frac{45}{10} \rightarrow \text{part} = 4.5$

PRACTICES:

Calculate the percentages.

1) 75% of 45	2) 50% of 66
3) 90% of 58	4) 25% of 88
5) 5% of 100	6) 80% of 60

Solve.

7) What percentage of 60 is 6	8) 6.76 is what percentage of 52?
9) 17 is what percentage of 85?	10) Find what percentage of 96 is 24.

Score: ....

Answer Key

1) 33.75	2) 33
3) 52.2	4) 22
5) 5	6) 48
7) 10%	8) 13%
9) 20%	10) 25%

Name: ....

Percent Problems

- ✓ In each percent question, we are finding the base, or part or the percent.
- ✓ Use the following equations to find each missing portion.
  - Base = Part ÷ Percent
  - Part = Percent × Base
  - Percent = Part ÷ Base

EXAMPLE:

20 is 5% of what number?

Use the formula: $Base = Part \div Percent \rightarrow Base = 20 \div 0.05 = 400$

20 is 5% of 400

PRACTICES:

Solve each problem.

1) 52% of what number is 13?	2) What is 15% of 9 inches?
3) What percent of 185.6 is 23.2?	4) 24 is 72% of what?
5) 35 is what percent of 70?	6) 10 is 200% of what?
7) 14 is what percent of 70?	8) 26% of 100 is what number?

9) Mia requires 50% to pass. If she gets 250 marks and falls short by 90 marks, what were the maximum marks she could have got?

10) Jack scored 14 out of 70 marks in mathematics, 9 out of 10 marks in history and 56 out of 100 marks in science. In which subject his percentage of marks is the best?

Score: ....

Answer Key

1) 25	2) 1.35
3) 12.5	4) 33.33
5) 50%	6) 5
7) 20%	8) 26
9) 680	10) History

Name: ....

Markup, Discount, and Tax

- ✓ Markup = selling price − cost
- ✓ Markup rate = markup is divided by the cost
- ✓ Discount = Multiply the rate of discount by regular price.
- ✓ Tax: To find tax, multiply the taxable amount (income, property value, etc.) to the tax rate.
- ✓ To find tip, multiply selling price to the rate.

EXAMPLE:

With an **10%** discount, Ella was able to save **$20** on a dress. What was the original price of the dress?

$10\% \text{ of } x = 20, \frac{10}{100} \times x = 20, x = \frac{100 \times 20}{10} = 200$

PRACTICES:

Find the selling price of each item.

1) Cost of a chair: $20, markup: 30%, discount: 10%, tax: 10%

2) Cost of computer: $1,600.00, markup: 65%

3) Cost of a pen: $3.20, markup: 50%, discount: 15%, tax: 5%

4) Cost of a puppy: $1,800, markup: 40%, discount: 10%

5) Cost of a book: $50, markup: 40%, discount: 20%, tax: 5%

6) Original price of a tablet: $400, discount: 20% Tax: 5%,

7) Original price of a book: $50, markup:20% Discount: 20%, Tax: 2.5%,

8) Original price of a cellphone: $500, markup:14% Discount: 25%, Tax: 1.6%,

9) Original price of a sofa: $800, markup:10% Discount: 15%, Tax: 1.5%,

10) Original price of a car: $40,000, markup:12% Discount: 25%, Tax: 6.5%,

Score: ....

Answer Key

1) $25.74	2) $2,640
3) $4.284	4) $2,268
5) $58.8	6) $336
7) $49.2	8) $434.34
9) $759.22	10) $35,784

Simple Interest

- ✓ Simple Interest: The charge for borrowing money or the return for lending it.
  To solve a simple interest problem, use this formula:
- ✓ Interest = principal × rate × time ⇒ $I = p \times r \times t$

EXAMPLE:

Find simple interest for $5, 200 at 4% for 3 years.

Use Interest formula: $I = prt$

$P = \$5{,}200$, $r = 4\% = \frac{4}{100} = 0.04$ and $t = 3$

Then: $I = 5{,}200 \times 0.04 \times 3 = \$624$

PRACTICES:

Use simple interest to find the ending balance.

1) $1,200 at 15% for 3 years.	2) $320,000 at 2.85% for 7 years.
3) $1,500 at 2.25% for 12 years.	4) $12,500 at 6.2% for 4 years.
5) $31,000 at 1.5% for 10 months.	6) $18,000 at 5.2% for 5 years.

7) Emily puts $6,000 into an investment yielding 3.25% annual simple interest; she left the money in for 3 years. How much interest does Sara get at the end of those 3 years?

8) A new car, valued at $42,000, depreciates at 7.5% per year from original price. Find the value of the car 6 years after purchase.

9) $880 interest is earned on a principal of $2,200 at a simple interest rate of 4% interest per year. For how many years was the principal invested?

10) A bank is offering 3.2% simple interest on a savings account. If you deposit $15,000, how much interest will you earn in six years?

Score: ....

Answer Key

1) $1,740	2) $383,840.00
3) $1,905.00	4) $15,600
5) $31,387.50	6) $22,680
7) $585.00	8) $23,100
9) 10 years	10) $2,880

TASC Math Prep

Name: ....

Converting Between Percent, Fractions, and Decimals

- ✓ To a percent: We move the decimal point 2 places to the right and add the percentage (%) symbol.
- ✓ Divide by 100 to change a number from percent to decimal.

EXAMPLE:

$30\% = 0.30$
$0.24 = 24\%$

PRACTICES:

Converting fractions to decimals.	Write each decimal as a percent.
1) $\frac{23}{10}$	2) 0.002
3) $\frac{2}{20}$	4) 0.08
5) $\frac{7}{100}$	6) 0.2
7) $\frac{20}{50}$	8) 3.25
9) $\frac{3}{60}$	10) 1.01

Score: ....

Answer Key

1) 2.3	2) 0.2%
3) 0.1	4) 8%
5) 0.07	6) 20%
7) 0.4	8) 325%
9) 0.05	10) 101%

Chapter 4 : Exponents and Radicals

Topics that you'll learn in this chapter:

- Multiplication Property of Exponents
- Division Property of Exponents
- Powers of Products and Quotients
- Zero, Negative Exponents and Bases

"Mathematics is no more computation than typing is literature." – *John Allen Paulos*

Name: ....

Multiplication Property of Exponents

- ✓ Exponents are shorthand for recurrent multiplication of the identical number by itself. For example, instead of writing 2×2, we can write 2^2. For $3 \times 3 \times 3 \times 3$, we can write 3^4
- ✓ In algebra, a variable is a letter used as a replacement for a number. The most common letters are: $x, y, z, a, b, c, m,$ and n.
- ✓ Exponent's rules: $(x^a)^b = x^{a \times b}$, $\qquad (xy)^a = x^a \times y^a$,

    $x^a \times x^b = x^{a+b}$, $\qquad x^a \times y^a = (xy)^a$,

EXAMPLE:

Multiply. $-2x^5 \times 7x^3 =$

Use Exponent's rules: $x^a \times x^b = x^{a+b} \rightarrow x^5 \times x^3 = x^{5+3} = x^8$

Then: $-2x^5 \times 7x^3 = -14x^8$

PRACTICES:

Simplify.

1) $4^3 \times 4^2$	2) $2 \times 2^2 \times 2^3$
3) $2^4 \times 2$	4) $8x^2 \times x$
5) $15x^7 \times x$	6) $3x \times x^3$
7) $2x^5 \times 5x^4$	8) $5x^2 \times 3x^2y^2$
9) $6y^5 \times 8xy^2$	10) $5xy^3 \times 4x^3y^2$

Answer Key

1) 4^5	2) 2^6
3) 2^5	4) $8x^3$
5) $15x^8$	6) $3x^4$
7) $10x^9$	8) $15x^4y^2$
9) $48xy^7$	10) $20x^4y^5$

Division Property of Exponents

- For division of exponents, we can use these formulas: $\left(\frac{a}{b}\right)^c = \frac{a^c}{b^c}$, $b \neq 0$

$$\frac{x^a}{x^b} = x^{a-b}, x \neq 0 \qquad \frac{x^a}{y^a} = \left(\frac{x}{y}\right)^a, y \neq 0$$

$$\frac{x^a}{x^b} = \frac{1}{x^{b-a}}, x \neq 0, \qquad \frac{1}{x^b} = x^{-b}$$

EXAMPLE:

Simplify. $\frac{4x^3y}{36x^2y^3} =$

First you cancel the common factor: $4 \rightarrow \frac{4x^3y}{36x^2y^3} = \frac{x^3y}{9x^2y^3}$

Use Exponent's rules: $\frac{x^a}{x^b} = x^{a-b} \rightarrow \frac{x^3}{x^2} = x^{3-2}$

Then: $\frac{4x^3y}{36x^2y^3} = \frac{xy}{9y^3} \rightarrow$ now cancel the common factor: $y \rightarrow \frac{xy}{9y^3} = \frac{x}{9y^2}$

PRACTICES:

Simplify.

1) $\frac{4^3}{4}$

2) $\frac{51}{51^{14}}$

3) $\frac{5^2}{5^3}$

4) $\frac{3^4}{15^4}$

5) $\frac{x}{x^7}$

6) $\frac{42x^2}{6x^2}$

7) $\frac{3x^{-3}}{12x^{-1}}$

8) $\frac{81x^5}{9x^3}$

9) $\frac{3x^4}{4x^5}$

10) $\frac{21x}{3x^2}$

Score: ....

Answer Key

1) 4^2	2) $\frac{1}{51^{13}}$
3) $\frac{1}{5}$	4) $(\frac{1}{5})^4 = \frac{1}{5^4}$
5) $\frac{1}{x^6}$	6) 7
7) $\frac{1}{4x^2}$	8) $9x^2$
9) $\frac{3}{4x}$	10) $\frac{7}{x}$

Powers of Products and Quotients

- For any number except zero, a and b and any integer x, $(ab)^x = a^x \times b^x$.

EXAMPLE:

Simplify. $(3x^5y^4)^2 =$

Use Exponent's rules: $(x^a)^b = x^{a \times b}$

$$(3x^5y^4)^2 = (3)^2(x^5)^2(y^4)^2 = 9x^{5 \times 2}y^{4 \times 2} = 9x^{10}y^8$$

PRACTICES:

Simplify.

1) $(5x^3)^2$

2) $(xy)^2$

3) $(ax^2)^3$

4) $(2x^3yz)^2$

5) $(4x^2y^3)^2$

6) $(5x^2y^3)^2$

7) $(2xy^2)^3$

8) $(2x^3y)^4$

9) $(7x^4y^8)^2$

10) $(10x)^3$

TASC Math Prep

Score: ....

Answer Key

1) $25x^6$	2) x^2y^2
3) a^3x^6	4) $4x^6y^2z^2$
5) $16x^4y^6$	6) $25x^4y^6$
7) $8x^3y^6$	8) $8x^{12}y^4$
9) $49x^8y^{16}$	10) $1,000x^3$

Zero and Negative Exponents

- A negative exponent just means that the base is on the wrong side of the fraction line, so you need to flip the base to the other side. For example, "x^{-2}" (pronounced as "ecks to the minus two") just means "x^2" but below, as in $\frac{1}{x^2}$

EXAMPLE:

Evaluate. $\left(\frac{4}{9}\right)^{-2} =$

Use Exponent's rules: $\frac{1}{x^b} = x^{-b} \rightarrow \left(\frac{4}{9}\right)^{-2} = \frac{1}{\left(\frac{4}{9}\right)^2} = \frac{1}{\frac{4^2}{9^2}}$

Now use fraction rule: $\frac{1}{\frac{b}{c}} = \frac{c}{b} \rightarrow \frac{1}{\frac{4^2}{9^2}} = \frac{9^2}{4^2} = \frac{81}{16}$

PRACTICES:

Evaluate the following expressions.

1) 4^{-2}	2) 5^{-2}
3) 6^{-2}	4) 3^{-4}
5) 10^{-1}	6) 33^{-1}
7) 6^{-1}	8) 3^{-2}
9) 9^{-2}	10) 4^{-1}

Score: ....

Answer Key

1) $\frac{1}{16}$	2) $\frac{1}{25}$
3) $\frac{1}{36}$	4) $\frac{1}{81}$
5) $\frac{1}{10}$	6) $\frac{1}{33}$
7) $\frac{1}{6}$	8) $\frac{1}{9}$
9) $\frac{1}{81}$	10) $\frac{1}{4}$

Negative Exponents and Negative Bases

- ✓ First make the power positive. A negative exponent can be written as reciprocal of that number with a positive exponent.
- ✓ The parenthesis is significant!
- ✓ 5^{-3} is not the same as $(-5)^{-3}$

$$(-5)^{-3} = -\frac{1}{5^3} \text{ and } (5)^{-3} = +\frac{1}{5^3}$$

EXAMPLE:

Simplify. $\left(-\frac{5x}{3yz}\right)^{-3} =$

Use Exponent's rules: $\frac{1}{x^b} = x^{-b} \rightarrow \left(-\frac{5x}{3yz}\right)^{-3} = \frac{1}{\left(-\frac{5x}{3yz}\right)^3} = \frac{1}{-\frac{5^3 x^3}{3^3 y^3 z^3}}$

Now use fraction rule: $\frac{1}{\frac{b}{c}} = \frac{c}{b} \rightarrow \frac{1}{-\frac{5^3 x^3}{3^3 y^3 z^3}} = -\frac{3^3 y^3 z^3}{5^3 x^3} = -\frac{27 y^3 z^3}{125 x^3}$

PRACTICES:

Simplify.

1) 7^{-1}	2) $-2x^{-2}$
3) $\frac{x}{x^{-5}}$	4) $-\frac{a^{-2}}{b^{-1}}$
5) $\frac{7}{x^{-5}}$	6) $\frac{2b}{-5c^{-2}}$
7) $\frac{2n^{-1}}{12p^{-2}}$	8) $\frac{8b^{-4}}{3c^{-2}}$
9) $89xy^{-2}$	10) $\left(\frac{1}{3}\right)^{-2}$

TASC Math Prep

Score: ....

Answer Key

1) $\dfrac{1}{7}$	2) $-\dfrac{2}{x^2}$
3) x^5	4) $-\dfrac{b^1}{a^2}$
5) $7x^5$	6) $-2\dfrac{bc^2}{5}$
7) $\dfrac{p^2}{6n}$	8) $\dfrac{8c^2}{3b^4}$
9) $\dfrac{89x}{y^2}$	10) 9

Writing Scientific Notation

- ✓ It is used to write very big or very small values in decimal representation.
- ✓ In scientific notation all numbers can be written in the form of:

$m \times 10^n$, where $1 \leq m \leq 10$ and n is any integer.

Decimal notation	Scientific notation
5	5×10^0
$-25{,}000$	-2.5×10^4
0.5	5×10^{-1}
2,122.456	2.122456×10^3

EXAMPLE:

Write 8.3×10^{-5} in standard notation.

$10^{-5} \rightarrow$ When the decimal moved to the right, the exponent is negative.

Then: $8.3 \times 10^{-5} = 0.000083$

PRACTICES:

Write each number in scientific notation.

1) 12,000,000	2) 25×10^5
3) 0.0015	4) 54,000
5) 0.0005021	6) 666,012
7) 0.00000076	8) 102,900,000
9) 4,100,000,000	10) 3,600,000

Score: ....

Answer Key

1) 1.2×10^7	2) 2.5×10^6
3) 1.5×10^{-3}	4) 5.4×10^4
5) 5.021×10^{-4}	6) 6.66012×10^5
7) 7.6×10^{-7}	8) 1.029×10^8
9) 4.1×10^9	10) 3.6×10^6

Name: ....

Square Roots

- A square root of x is a number p whose square is: $p^2 = x$

    p is a square root of x.

EXAMPLE:

Find the square root of $\sqrt{225}$.

First factor the number: $225 = 15^2$, Then: $\sqrt{225} = \sqrt{15^2}$

Now use radical rule: $\sqrt[n]{a^n} = a$

Then: $\sqrt{15^2} = 15$

PRACTICES:

Find the value each square root.

1) $\sqrt{25}$	2) $\sqrt{1,600}$
3) $\sqrt{100}$	4) $\sqrt{121}$
5) $\sqrt{4}$	6) $\sqrt{225}$
7) $\sqrt{10,000}$	8) $\sqrt{16}$
9) $\sqrt{64}$	10) $\sqrt{36}$

Score: ....

Answer Key

1) 5	2) 40
3) 10	4) 11
5) 2	6) 15
7) 100	8) 4
9) 8	10) 6

Chapter 5 : Algebraic Expressions

Topics that you'll learn in this chapter:

- Expressions and Variables
- Simplifying Variable and Polynomial Expressions
- Translate Phrases into an Algebraic Statement
- The Distributive Property
- Evaluating One and two Variable
- Combining like Terms

"Without mathematics, there's nothing you can do. Everything around you are mathematics. Everything around you are numbers." – Shakuntala Devi

Name: ....

Translate Phrases into an Algebraic Statement

How to translate key words and phrases into algebraic expressions:

- ✓ Addition: the sum of, more than, plus, etc.
- ✓ Subtraction: less than, decreased, minus, etc.
- ✓ Multiplication: times, multiplied, product, etc.
- ✓ Division: quotient, ratio, divided, etc.

EXAMPLE:

5 times the sum of x and 8

Sum of 8 and x: $8 + x$. Times is used for multiplication. Then: $5 \times (8 + x)$

PRACTICES:

Write an algebraic expression for each phrase.

1) Fifteen subtracted from a number.

2) The quotient of seventeen and a number.

3) A number increased by fifty.

4) A number divided by -21.

5) The difference between sixty-three and a number.

6) Threefold a number decreased by 45.

7) Seven times the sum of a number and -21.

8) The quotient of 90 and the product of a number and -8.

9) Nine subtracted from 4 times a number.

10) The difference of six and a number.

Score: ....

Answer Key

1) $x - 15$	2) $\dfrac{17}{x}$
3) $x + 50$	4) $-\dfrac{x}{21}$
5) $63 - x$	6) $3x - 45$
7) $7(x + (-21))$	8) $-\dfrac{90}{8x}$
9) $4x - 9$	10) $6 - x$

The Distributive Property

- Distributive Property:

$$a(b + c) = ab + ac$$

EXAMPLE:

Simply. $(5x - 3)(-5) =$

Use Distributive Property formula: $a(b + c) = ab + ac$

$$(5x - 3)(-5) = -25x + 15$$

PRACTICES:

Use the distributive property to simply each expression.

1) $4(9 - 3x)$

2) $-(-8 - 4x)$

3) $(-5x - 1)(-2)$

4) $(-3)(2x - 4)$

5) $4(5 + 3x)$

6) $(-9x + 10)3$

7) $(-4 - 5x)(-3)$

8) $(-2x)(-3 + 2x) - 3x(1 - 5x)$

9) $(-2)(3x - 1) + 4(3x + 2)$

10) $(-15)(2x + 3)$

Score: ....

Answer Key

1) $-12x + 36$	2) $4x + 8$
3) $10x + 2$	4) $-6x + 12$
5) $12x + 20$	6) $-27x + 30$
7) $15x + 12$	8) $11x^2 + 3x$
9) $6x + 10$	10) $-30x - 45$

Evaluating One Variable

- To solve a variable expression, find the variable and substitute a number for that variable.
- Perform the mathematical operations.

EXAMPLE:

Solve this expression. $12 - 2x, x = -1$

First substitute -1 for x, then:

$$12 - 2x = 12 - 2(-1) = 12 + 2 = 14$$

PRACTICES:

Simplify each algebraic expression.

1) $5x + 4, x = 1$

2) $x + (-4), x = -6$

3) $-10x + 8, x = -2$

4) $\left(-\frac{36}{x}\right) - 10 + 2x, x = 6$

5) $\frac{36}{x} - 3, x = 3$

6) $(-10) - \frac{x}{4} + 4x, x = -8$

7) $15 + 6x - 3, x = -1$

8) $(-5) + \frac{x}{8}, x = 64$

9) $\left(-\frac{24}{x}\right) - 10 + 5x, x = 4$

10) $(-4) + \frac{4x}{9}, x = 81$

Score: ....

Answer Key	
1) 9	2) −10
3) 28	4) −4
5) 9	6) −40
7) 6	8) 3
9) 4	10) 32

Evaluating Two Variables

To solve an algebraic expression, substitute a number for each variable and perform the mathematical operations.

EXAMPLE:

Solve this expression. $-3x + 5y$, $x = 2, y = -1$

First substitute 2 for x, and -1 for y, then:
$$-3x + 5y = -3(2) + 5(-1) = -6 - 5 = -11$$

PRACTICES:

Simplify each algebraic expression.

1) $5a - (5 - b)$,
   $a = 2, b = 3$

2) $5x + 3y - 6 + 3y$,
   $x = 3, y = 1$

3) $\left(-\frac{27}{x}\right) + 4 + 3y$,
   $x = 3, y = 5$

4) $(-4)(-3a - 5b)$,
   $a = 3, b = 4$

5) $7x + 10 - 5y$,
   $x = 3, y = 6$

6) $18 + 3(-x - 4y)$,
   $x = 2, y = 5$

7) $12x + 2y$,
   $x = 5, y = 10$

8) $x \times 6 \div 3y$,
   $x = 6, y = 1$

9) $4x - 3y$,
   $x = 6, y = 3$

10) $\left(-\frac{14}{x}\right) + 4y$,
    $x = 7, y = -3$

Score: ....

Answer Key

1) 8	2) 15
3) 10	4) 116
5) 1	6) −48
7) 80	8) 12
9) 15	10) −14

Name: ....

Expressions and Variables

- In algebra, a variable is a letter used as a replacement for a number. The most common letters are: $x, y, z, a, b, c, m,$ and n.
- An algebraic expression is an expression that has variables, integers, and math operations such as addition, subtraction, multiplication, division, etc.
- In an expression, we can combine "identical" terms. (Values with same power and variable)

EXAMPLE:

Simplify this expression. $(10x + 2x + 3) = ?$

Combine like terms. Then: $(10x + 2x + 3) = 12x + 3$ (remember you cannot combine variables and numbers

PRACTICES:

Simplify each expression.

1) $10(-3 - 8x), x = 4$	2) $-3(5 - 8x) - 6x, x = 1$
3) $2x - 8x, x = 2$	4) $x + 12x, x = 6$
5) $20 - 5x + 10x + 5, x = 3$	6) $15(5x + 3), x = 0$
7) $20(4 - x) - 9, x = 2$	8) $20x - 8x - 10, x = 5$
9) $6x + 9y, x = 4, y = 2$	10) $6x - 2x, x = 8$

Score: ....

Answer Key

1) −350	2) 3
3) −12	4) 78
5) 40	6) 45
7) 31	8) 50
9) 42	10) 32

Combining like Terms

- ✓ We separate the terms by "+" and "−" signs.
- ✓ Identical terms are those terms with same powers and same variables.
  Make sure to use the "+" or "−" that is in front of the coefficient

EXAMPLE:

Simplify this expression. $(-5)(8x - 6) =$

Use Distributive Property formula: $a(b + c) = a + ac$

$(-5)(8x - 6) = -40x + 30$

PRACTICES:

Simplify each expression.

1) $-8(-5x + 1)$

2) $6(-2 + 4x)$

3) $-8 - 14x + 16x + 3$

4) $9x - 7x - 15 + 18$

5) $(-9)(12x - 21) + 31$

6) $2(4x + 9) + 12x$

7) $4(-2x - 17) + 14(3x + 1)$

8) $(9x - 5y)7 + 25y$

9) $4.5x^3 \times (-8x)$

10) $-19 - 15x^2 + 12x^2$

Score: ....

Answer Key

1) $40x - 8$	2) $24x - 12$
3) $2x - 5$	4) $2x + 3$
5) $220 - 108x$	6) $20x + 18$
7) $34x - 54$	8) $63x - 10y$
9) $-36x^4$	10) $-3x^2 - 19$

Simplifying Polynomial Expressions

- A polynomial is a unique expression that consists of coefficients and variables that performs only the arithmetic of addition, subtraction, multiplication, and non-negative integer exponents of variables.

$$P(x) = a_n x^n + a_{n-1} x^{n-1} + \ldots + a_2 x^2 + a_1 x + a_0$$

EXAMPLE:

Simplify this Polynomial Expressions. $4x^2 - 5x^3 + 15x^4 - 12x^3 =$

Combine "like" terms: $-5x^3 - 12x^3 = -17x^3$

Then: $4x^2 - 5x^3 + 15x^4 - 12x^3 = 4x^2 - 17x^3 + 15x^4$

Then write in standard form: $4x^2 - 17x^3 + 15x^4 = 15x^4 - 17x^3 + 4x^2$

PRACTICES:

Simplify each polynomial.

1) $(2x^2 + 4) - (9 + 5x^2)$	2) $(25x^3 - 12x^2) - (6x^2 - 9x^3)$
3) $14x^5 - 15x^6 + 2x^5 - 16x^6 + x^6$	4) $(15 + 12x^3) + (3x^3 + 5)$
5) $13x^3 - 15x^4 + 12x^3 + 20x^4$	6) $-6x^2 + 15x^2 + 17x^3 + 16 - 32$
7) $15x^3 + 12 + 2x^2 - 5x - 10x$	8) $24x^2 - 16x^3 - 4x(2x^2 + 3x)$
9) $(21x^4 - 10x) - (2x - x^4)$	10) $(7x^2 - 9) + (x^2 - 8x^3)$

Score: ....

Answer Key

1) $-3x^2 - 5$	2) $34x^3 - 18x^2$
3) $-30x^6 + 16x^5$	4) $15x^3 + 20$
5) $5x^4 + 25x^3$	6) $17x^3 + 9x^2 - 16$
7) $15x^3 + 2x^2 - 15x + 12$	8) $-24x^3 + 12x^2$
9) $22x^4 - 12x$	10) $-8x^3 + 8x^2 - 9$

Chapter 6 : Equations and Inequalities

Topics that you'll learn in this chapter:

- One, Two, and Multi – Step Equations
- Graphing Single– Variable Inequalities
- One, Two, and Multi – Step Inequalities
- Solving Systems of Equations by Substitution and Elimination
- Finding Slope and Writing Linear Equations
- Graphing Lines Using Slope– Intercept and Standard Form
- Graphing Linear Inequalities
- Finding Midpoint and Distance of Two Points

"The study of mathematics, like the Nile, begins in minuteness but ends in magnificence." – *Charles Caleb Colton*

One–Step Equations

- ✓ The values of two algebraic expressions on both sides of an equation are always equal.

$$ax + b = c$$

- ✓ You only require performing one Math operation to solve the problem.

EXAMPLE:

Solve this equation. $x + 24 = 0, x = ?$

Here, we have the addition operation, and its inverse operation is subtraction. To solve this equation, subtract 24 from both sides of the equation: $x + 24 - 24 = 0 - 24$

Then simplify: $x + 24 - 24 = 0 - 24 \rightarrow x = -24$

PRACTICES:

Solve each equation.

1) $x + 4 = 16$	2) $48 = (-2) + x$
3) $5x = (-105)$	4) $(-8) = (8x)$
5) $(-2) = 14 + x$	6) $5 + x = 6$
7) $2x + 3 = (-7)$	8) $28 = x + 7$
9) $(-15) + x = (-15)$	10) $12x = (-36)$

Score: ....

Answer Key

1) 12	2) 50
3) −21	4) −1
5) −16	6) 1
7) −5	8) 21
9) 0	10) −3

Two-Step Equations

- ✓ You only require performing two math operations (add, subtract, multiply, or divide) to solve the equation.
- ✓ Simplify the equation using the inverse of addition or subtraction.
- ✓ Simplify the equation further by using the inverse of division or multiplication.

EXAMPLE:

Solve this equation. $3x = 15, x = ?$

Here, we have the multiplication operation (variable x is multiplied by 3) and its inverse operation is division. To solve this problem, we divide both sides of equation by 3:

$3x = 15 \rightarrow 3x \div 3 = 15 \div 3 \rightarrow x = 5$

PRACTICES:

Solve each equation.

1) $4(2 + 2x) = 8$	2) $(-5)(x - 3) = 25$
3) $(-5)(2x - 5) = (-15)$	4) $4(9 + 3x) = -12$
5) $6(2x + 1) = 30$	6) $2(x + 2) = 42$
7) $2(12 + 6x) = 60$	8) $(-10)(5x) = 100$
9) $4(3x + 3) = 24$	10) $\dfrac{x - 5}{3} = 4$

Score: ....

Answer Key

1) 0	2) −2
3) 4	4) −4
5) 2	6) 10
7) 3	8) −2
9) 1	10) 17

Name: ....

Multi–Step Equations

- ✓ Combine "identical" terms on one side.
- ✓ Put variables to one side by adding or subtracting.
- ✓ Simplify the equation by using the inverse of addition or subtraction.
- ✓ Simplify further by using the inverse of division or multiplication.

EXAMPLE:

Solve this equation. $-(2 - x) = 5$

First, use Distributive Property: $-(2 - x) = -2 + x$

Now by adding 2 to both sides of the equation, we can solve it.

$-2 + x = 5 \rightarrow -2 + x + 2 = 5 + 2$
Now simplify: $-2 + x + 2 = 5 + 2 \rightarrow x = 7$

PRACTICES:

Solve each equation.

1) $8 - 2x = 28$	2) $-10 = -(x + 7)$
3) $2x - 17 = (-x) + 1$	4) $-2x = (-3x) - 8$
5) $5(14 + 2x) + 3x = -x$	6) $x - 11 = x - 5 + 2x$
7) $15 + 2x = (-25) - 2x + 3x$	8) $-3(x - 3x) = 40 - 4x$
9) $24 + 8x + x = (-x + 4)$	10) $-8(1 + 5x) = 152$

Score: ....

Answer Key

1) −10	2) 3
3) 6	4) −8
5) −5	6) −3
7) −40	8) 4
9) −2	10) −4

Graphing Single–Variable Inequalities

- ✓ Inequality is like equations and uses symbols for "less than" (<) and "greater than" (>).
- ✓ To solve inequalities, we need to separate the variable. (Like in equations)
- ✓ Find the value of the inequality on the number line to graph an inequality.
- ✓ For greater than or less than draw open circle on the value of the variable.
- ✓ Use filled circle if there is an equal sign too.
- ✓ Draw a line to the left or to the right for less or greater than.

EXAMPLE:

Draw a graph for $x > 2$

Since, the variable is greater than 2, then we need to find 2 and draw an open circle above it. Then, draw a line to the right.

PRACTICES:

Draw a graph for each inequality.

1) $2 \geq x$

2) $x < 3$

3) $5 \geq x$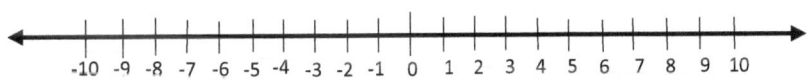

4) $x \geq -2$

5) $x > 0$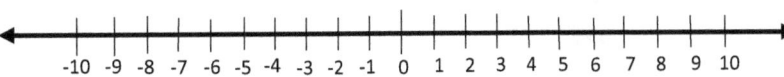

6) $-1.5 < x$		
7) $x \geq -1$		

Score: ....

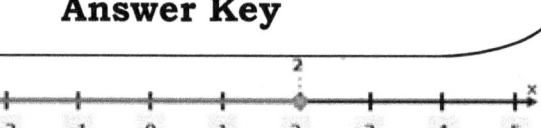

Answer Key

1) $2 \geq x$	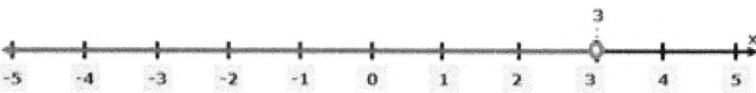	
2) $x < 3$	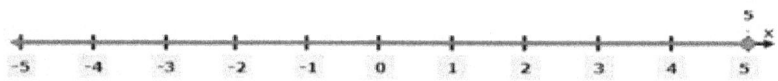	
3) $5 \geq x$	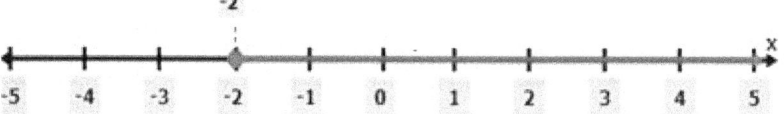	
4) $x \geq -2$	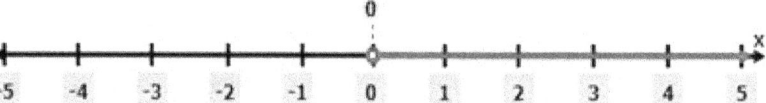	
5) $x > 0$		
6) $-1.5 < x$	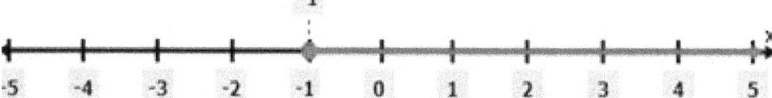	
7) $x \geq -1$		

One–Step Inequalities

- ✓ Like equations, first separate the variable by using inverse operation.
- ✓ For dividing or multiplying both sides by negative numbers, flip the direction of the inequality sign.

EXAMPLE:

Solve this inequality. $x - 1 \leq 2$

Add 1 to both sides. $x - 1 \leq 2 \rightarrow x - 1 + 1 \leq 2 + 1$, then: $x \leq 3$

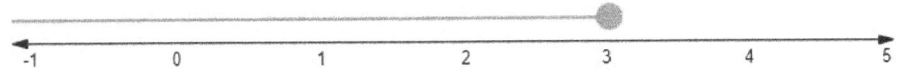

PRACTICES:

Solve each inequality and graph it.

1) $2x + 3 \geq 7$

2) $x < 3$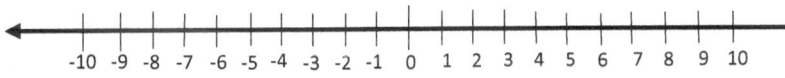

3) $5 \geq x$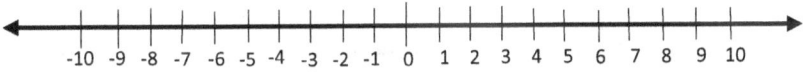

4) $x \geq -2$

5) $x > 0$

6) $-1.5 < x$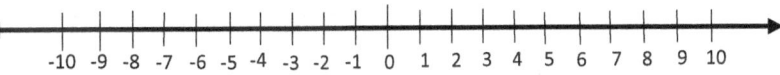

7) $x \geq -1$

Score: ....

Answer Key

1) $2 \geq x$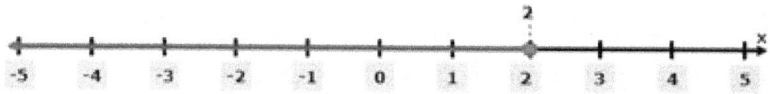

2) $x < 3$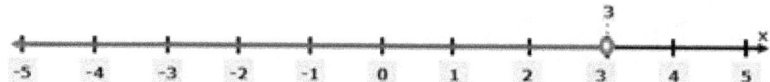

3) $5 \geq x$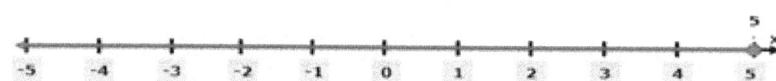

4) $x \geq -2$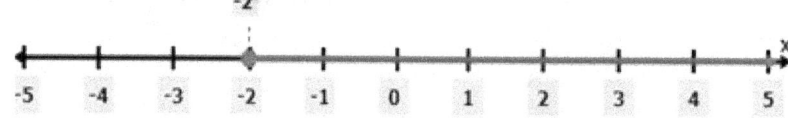

5) $x > 0$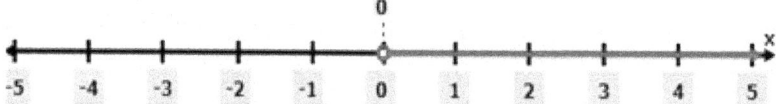

6) $-1.5 < x$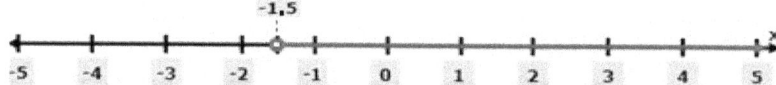

7) $x \geq -1$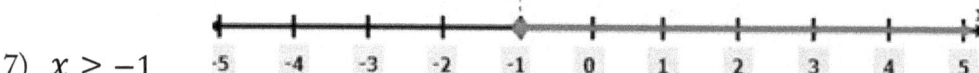

Two-Step Inequalities

- ✓ Separate the variable.
- ✓ Flip the direction of the inequality sign for dividing both sides by negatives numbers.
- ✓ We can simplify by using the inverse of addition or subtraction.
- ✓ We can simplify further by using the inverse of division or multiplication.

EXAMPLE:

Solve: $2x + 9 \geq 11$

First add -9 to both sides: $2x + 9 - 9 \geq 11 - 9 \rightarrow 2x \geq 2$

Now, divide both sides by 2: $2x \geq 2 \rightarrow x \geq 1$

PRACTICES:

Solve each inequality and graph it.

1) $x - 4 \leq 4$	2) $x + 4 \geq 5$
3) $3x - 2 \leq 7$	4) $5x + 2 < 12$
5) $x + 7 \geq 9$	6) $3x - 3 \leq 3$
7) $7x - 4 < 3$	8) $8 + x \leq 13$
9) $2x + 7 \leq 11$	10) $10x - 16 < 4$

Answer Key

1) $x \leq 8$	2) $x \geq 1$
3) $x \leq 3$	4) $x < 2$
5) $x \geq 2$	6) $x \leq 2$
7) $x < 1$	8) $x \leq 5$
9) $x \leq 2$	10) $x < 2$

Multi–Step Inequalities

- ✓ Separate the variable.
- ✓ We can simplify by using the inverse of addition or subtraction.
- ✓ We can simplify further by using the inverse of division or multiplication.

EXAMPLE:

Solve this inequality. $2x - 2 \leq 6$

First add 2 to both sides: $2x - 2 + 2 \leq 6 + 2 \rightarrow 2x \leq 8$

Now, divide both sides by 2: $2x \leq 8 \rightarrow x \leq 4$

PRACTICES:

Solve each inequality.

1) $-(x + 3) + 8 < 25$

2) $\frac{3x + 1}{2} \leq 5$

3) $\frac{x - 4}{3} > 7$

4) $4(x - 2) \leq 8$

5) $\frac{x}{3} + \frac{1}{3} < 2$

6) $\frac{x+4}{5} > 3$

7) $\frac{x}{8} + \frac{3}{4} < 1$

8) $2(x + 5) + 4 > 10$

9) $24 + 5x < 4$

10) $\frac{x+2}{3} > 10$

Score: ....

Answer Key

1) $x > -20$	2) $x \leq 3$
3) $x > 25$	4) $x \leq 4$
5) $x < 5$	6) $x > 11$
7) $x < 2$	8) $x > -2$
9) $x < -4$	10) $x > 28$

Solving Systems of Equations by Substitution

✓ Let the system of equations. $x + y = 1; -2x + y = 4$

Put $x = 1 - y$ in the second equation.

$-2(1 - y) + y = 4 \rightarrow -2 + 2y + y = 4 \Rightarrow y = 2$

Put $y = 2$ in $x = 1 - y$; then $x = 1 - 2 = -1$; $(-1, 2)$

EXAMPLE:

Solve: $-2x - 2y = -13; -4x + 2y = 10$

For the first equation above, you can add $-4x + 2y$ to the left side and 10 to the right side of the first equation: $-2x - 2y + (-4x + 2y) = -13 + 10$. Now, if you simplify, you get: $-2x - 2y - 4x + 2y = -3 \rightarrow -6x = -3 \rightarrow$ $x = 0.5$. Now, put 0.5 for the x in the first equation: $-2(0.5) - 2y = -13$. By solving this equation, $y = 6$

PRACTICES:

Solve each system of equation by substitution.

1) $-x + 5y = -4$ $x - 3y = 8$	2) $2x + 3y = -6$ $-2x - y = 8$
3) $x + 2y = -5$ $5x - 10y = 5$	4) $y = -x + 5$ $3x - y = -3$
5) $3x = 6$ $10y = 4x + 2$	6) $3x + 2y = 2$ $x + 4y = -6$
7) $4x + y = 3$ $2x + 4y = -2$	8) $4y = 2x + 3$ $x - 4y = -2$
9) $7y = 14x$ $2x - 5y = -24$	10) $5y = x + 2$ $3x - 12y = -5$

Score: ....

Answer Key

1) $(14, 2)$	2) $(-\frac{9}{2}, 1)$
3) $(-2, -\frac{3}{2})$	4) $(\frac{1}{2}, \frac{9}{2})$
5) $(2, 1)$	6) $(2, -2)$
7) $(1, -1)$	8) $(-1, \frac{1}{4})$
9) $(3, 6)$	10) $(-\frac{1}{3}, \frac{1}{3})$

Multiplication Property of Exponents

- Exponents are shorthand for recurrent multiplication of the identical number by itself. For example, instead of writing 2×2, we can write 2^2. For $3 \times 3 \times 3 \times 3$, we can write 3^4
- In algebra, a variable is a letter used as a replacement for a number. The most common letters are: $x, y, z, a, b, c, m,$ and n.
- Exponent's rules: $(x^a)^b = x^{a \times b}$, $\quad (xy)^a = x^a \times y^a$,
  $x^a \times x^b = x^{a+b}$, $\quad x^a \times y^a = (xy)^a$,

EXAMPLE:

Multiply. $-2x^5 \times 7x^3 =$

Use Exponent's rules: $x^a \times x^b = x^{a+b} \rightarrow x^5 \times x^3 = x^{5+3} = x^8$

Then: $-2x^5 \times 7x^3 = -14x^8$

PRACTICES:

Simplify.

1) $4^3 \times 4^2$	2) $2 \times 2^2 \times 2^3$
3) $2^4 \times 2$	4) $8x^2 \times x$
5) $15x^7 \times x$	6) $3x \times x^3$
7) $2x^5 \times 5x^4$	8) $5x^2 \times 3x^2y^2$
9) $6y^5 \times 8xy^2$	10) $5xy^3 \times 4x^3y^2$

Score: ....

Answer Key

1) 4^5	2) 2^6
3) 2^5	4) $8x^3$
5) $15x^8$	6) $3x^4$
7) $10x^9$	8) $15x^4y^2$
9) $48xy^7$	10) $20x^4y^5$

Division Property of Exponents

✓ For division of exponents, we can use these formulas: $\left(\frac{a}{b}\right)^c = \frac{a^c}{b^c}, b \neq 0$

$$\frac{x^a}{x^b} = x^{a-b}, x \neq 0 \qquad \frac{x^a}{y^a} = \left(\frac{x}{y}\right)^a, y \neq 0$$

$$\frac{x^a}{x^b} = \frac{1}{x^{b-a}}, x \neq 0, \qquad \frac{1}{x^b} = x^{-b}$$

EXAMPLE:

Simplify. $\frac{4x^3y}{36x^2y^3} =$

First you cancel the common factor: $4 \to \frac{4x^3y}{36x^2y^3} = \frac{x^3y}{9x^2y^3}$

Use Exponent's rules: $\frac{x^a}{x^b} = x^{a-b} \to \frac{x^3}{x^2} = x^{3-2}$

Then: $\frac{4x^3y}{36x^2y^3} = \frac{xy}{9y^3} \to$ now cancel the common factor: $y \to \frac{xy}{9y^3} = \frac{x}{9y^2}$

PRACTICES:

Simplify.

1) $\frac{4^3}{4}$

2) $\frac{51}{51^{14}}$

3) $\frac{5^2}{5^3}$

4) $\frac{3^4}{15^4}$

5) $\frac{x}{x^7}$

6) $\frac{42x^2}{6x^2}$

7) $\frac{3x^{-3}}{12x^{-1}}$

8) $\frac{81x^5}{9x^3}$

9) $\frac{3x^4}{4x^5}$

10) $\frac{21x}{3x^2}$

Score: ....

Answer Key

1) 4^2	2) $\dfrac{1}{51^{13}}$
3) $\dfrac{1}{5}$	4) $\left(\dfrac{1}{5}\right)^4 = \dfrac{1}{5^4}$
5) $\dfrac{1}{x^6}$	6) 7
7) $\dfrac{1}{4x^2}$	8) $9x^2$
9) $\dfrac{3}{4x}$	10) $\dfrac{7}{x}$

Powers of Products and Quotients

- For any number except zero, a and b and any integer x, $(ab)^x = a^x \times b^x$.

EXAMPLE:

Simplify. $(3x^5y^4)^2 =$

Use Exponent's rules: $(x^a)^b = x^{a \times b}$

$$(3x^5y^4)^2 = (3)^2(x^5)^2(y^4)^2 = 9x^{5\times 2}y^{4\times 2} = 9x^{10}y^8$$

PRACTICES:

Simplify.

1) $(5x^3)^2$

2) $(xy)^2$

3) $(ax^2)^3$

4) $(2x^3yz)^2$

5) $(4x^2y^3)^2$

6) $(5x^2y^3)^2$

7) $(2xy^2)^3$

8) $(2x^3y)^4$

9) $(7x^4y^8)^2$

10) $(10x)^3$

Score: ....

Answer Key	
1) $25x^6$	2) x^2y^2
3) a^3x^6	4) $4x^6y^2z^2$
5) $16x^4y^6$	6) $25x^4y^6$
7) $8x^3y^6$	8) $8x^{12}y^4$
9) $49x^8y^{16}$	10) $1{,}000x^3$

Zero and Negative Exponents

- A negative exponent just means that the base is on the wrong side of the fraction line, so you need to flip the base to the other side. For example, "x^{-2}" (pronounced as "ecks to the minus two") just means "x^2" but below, as in $\frac{1}{x^2}$

EXAMPLE:

Evaluate. $\left(\frac{4}{9}\right)^{-2} =$

Use Exponent's rules: $\frac{1}{x^b} = x^{-b} \rightarrow \left(\frac{4}{9}\right)^{-2} = \frac{1}{\left(\frac{4}{9}\right)^2} = \frac{1}{\frac{4^2}{9^2}}$

Now use fraction rule: $\frac{1}{\frac{b}{c}} = \frac{c}{b} \rightarrow \frac{1}{\frac{4^2}{9^2}} = \frac{9^2}{4^2} = \frac{81}{16}$

PRACTICES:

Evaluate the following expressions.

1) 4^{-2}	2) 5^{-2}
3) 6^{-2}	4) 3^{-4}
5) 10^{-1}	6) 33^{-1}
7) 6^{-1}	8) 3^{-2}
9) 9^{-2}	10) 4^{-1}

Score: ....

Answer Key

1) $\frac{1}{16}$	2) $\frac{1}{25}$
3) $\frac{1}{36}$	4) $\frac{1}{81}$
5) $\frac{1}{10}$	6) $\frac{1}{33}$
7) $\frac{1}{6}$	8) $\frac{1}{9}$
9) $\frac{1}{81}$	10) $\frac{1}{4}$

Negative Exponents and Negative Bases

- First make the power positive. A negative exponent can be written as reciprocal of that number with a positive exponent.
- The parenthesis is significant!
- 5^{-3} is not the same as $(-5)^{-3}$

$$(-5)^{-3} = -\frac{1}{5^3} \text{ and } (5)^{-3} = +\frac{1}{5^3}$$

EXAMPLE:

Simplify. $\left(-\frac{5x}{3yz}\right)^{-3} =$

Use Exponent's rules: $\frac{1}{x^b} = x^{-b} \rightarrow \left(-\frac{5x}{3yz}\right)^{-3} = \frac{1}{\left(-\frac{5x}{3yz}\right)^3} = \frac{1}{-\frac{5^3 x^3}{3^3 y^3 z^3}}$

Now use fraction rule: $\frac{1}{\frac{b}{c}} = \frac{c}{b} \rightarrow \frac{1}{\frac{5^3 x^3}{3^3 y^3 z^3}} = -\frac{3^3 y^3 z^3}{5^3 x^3} = -\frac{27 y^3 z^3}{125 x^3}$

PRACTICES:

Simplify.

1) 7^{-1}	2) $-2x^{-2}$
3) $\frac{x}{x^{-5}}$	4) $-\frac{a^{-2}}{b^{-1}}$
5) $\frac{7}{x^{-5}}$	6) $\frac{2b}{-5c^{-2}}$
7) $\frac{2n^{-1}}{12p^{-2}}$	8) $\frac{8b^{-4}}{3c^{-2}}$
9) $89xy^{-2}$	10) $\left(\frac{1}{3}\right)^{-2}$

Score: ....

Answer Key

1) $\dfrac{1}{7}$	2) $-\dfrac{2}{x^2}$
3) x^5	4) $-\dfrac{b^1}{a^2}$
5) $7x^5$	6) $-2\dfrac{bc^2}{5}$
7) $\dfrac{p^2}{6n}$	8) $\dfrac{8c^2}{3b^4}$
9) $\dfrac{89x}{y^2}$	10) 9

Name: ....

Writing Scientific Notation

✓ It is used to write very big or very small values in decimal representation.
✓ In scientific notation all numbers can be written in the form of:

$m \times 10^n$, where $1 \leq m \leq 10$ and n is any integer.

Decimal notation	Scientific notation
5	5×10^0
$-25{,}000$	-2.5×10^4
0.5	5×10^{-1}
2,122.456	2.122456×10^3

EXAMPLE:

Write 8.3×10^{-5} in standard notation.

$10^{-5} \rightarrow$ When the decimal moved to the right, the exponent is negative.

Then: $8.3 \times 10^{-5} = 0.000083$

PRACTICES:

Write each number in scientific notation.

1) 12,000,000	2) 25×10^5
3) 0.0015	4) 54,000
5) 0.0005021	6) 666,012
7) 0.00000076	8) 102,900,000
9) 4,100,000,000	10) 3,600,000

Score: ....

Answer Key

1) 1.2×10^7	2) 2.5×10^6
3) 1.5×10^{-3}	4) 5.4×10^4
5) 5.021×10^{-4}	6) 6.66012×10^5
7) 7.6×10^{-7}	8) 1.029×10^8
9) 4.1×10^9	10) 3.6×10^6

Name: ....

Square Roots

- A square root of x is a number p whose square is: $p^2 = x$

    p is a square root of x.

EXAMPLE:

Find the square root of $\sqrt{225}$.

First factor the number: $225 = 15^2$, Then: $\sqrt{225} = \sqrt{15^2}$

Now use radical rule: $\sqrt[n]{a^n} = a$

Then: $\sqrt{15^2} = 15$

PRACTICES:

Find the value each square root.

1) $\sqrt{25}$

2) $\sqrt{1,600}$

3) $\sqrt{100}$

4) $\sqrt{121}$

5) $\sqrt{4}$

6) $\sqrt{225}$

7) $\sqrt{10,000}$

8) $\sqrt{16}$

9) $\sqrt{64}$

10) $\sqrt{36}$

Score: ....

Answer Key

1) 5	2) 40
3) 10	4) 11
5) 2	6) 15
7) 100	8) 4
9) 8	10) 6

Chapter 5 : Algebraic Expressions

Topics that you'll learn in this chapter:

- Expressions and Variables
- Simplifying Variable and Polynomial Expressions
- Translate Phrases into an Algebraic Statement
- The Distributive Property
- Evaluating One and two Variable
- Combining like Terms

"Without mathematics, there's nothing you can do. Everything around you are mathematics. Everything around you are numbers." – Shakuntala Devi

Name: ....

Translate Phrases into an Algebraic Statement

How to translate key words and phrases into algebraic expressions:

- ✓ Addition: the sum of, more than, plus, etc.
- ✓ Subtraction: less than, decreased, minus, etc.
- ✓ Multiplication: times, multiplied, product, etc.
- ✓ Division: quotient, ratio, divided, etc.

EXAMPLE:

5 times the sum of x and 8

Sum of 8 and x: $8 + x$. Times is used for multiplication. Then: $5 \times (8 + x)$

PRACTICES:

Write an algebraic expression for each phrase.

1) Fifteen subtracted from a number.

2) The quotient of seventeen and a number.

3) A number increased by fifty.

4) A number divided by – 21.

5) The difference between sixty –three and a number.

6) Threefold a number decreased by 45.

7) Seven times the sum of a number and – 21.

8) The quotient of 90 and the product of a number and – 8.

9) Nine subtracted from 4 times a number.

10) The difference of six and a number.

Score: ....

Answer Key

1) $x - 15$	2) $\frac{17}{x}$
3) $x + 50$	4) $-\frac{x}{21}$
5) $63 - x$	6) $3x - 45$
7) $7(x + (-21))$	8) $-\frac{90}{8x}$
9) $4x - 9$	10) $6 - x$

The Distributive Property

✓ Distributive Property:
$$a(b + c) = ab + ac$$

EXAMPLE:

Simply. $(5x - 3)(-5) =$

Use Distributive Property formula: $a(b + c) = ab + ac$

$$(5x - 3)(-5) = -25x + 15$$

PRACTICES:

Use the distributive property to simply each expression.

1) $4(9 - 3x)$

2) $-(-8 - 4x)$

3) $(-5x - 1)(-2)$

4) $(-3)(2x - 4)$

5) $4(5 + 3x)$

6) $(-9x + 10)3$

7) $(-4 - 5x)(-3)$

8) $(-2x)(-3 + 2x) - 3x(1 - 5x)$

9) $(-2)(3x - 1) + 4(3x + 2)$

10) $(-15)(2x + 3)$

Answer Key

1) $-12x + 36$	2) $4x + 8$
3) $10x + 2$	4) $-6x + 12$
5) $12x + 20$	6) $-27x + 30$
7) $15x + 12$	8) $11x^2 + 3x$
9) $6x + 10$	10) $-30x - 45$

Evaluating One Variable

- ✓ To solve a variable expression, find the variable and substitute a number for that variable.
- ✓ Perform the mathematical operations.

EXAMPLE:

Solve this expression. $12 - 2x$, $x = -1$

First substitute -1 for x, then:

$$12 - 2x = 12 - 2(-1) = 12 + 2 = 14$$

PRACTICES:

Simplify each algebraic expression.

1) $5x + 4$, $x = 1$

2) $x + (-4)$, $x = -6$

3) $-10x + 8$, $x = -2$

4) $\left(-\frac{36}{x}\right) - 10 + 2x$, $x = 6$

5) $\frac{36}{x} - 3$, $x = 3$

6) $(-10) - \frac{x}{4} + 4x$, $x = -8$

7) $15 + 6x - 3$, $x = -1$

8) $(-5) + \frac{x}{8}$, $x = 64$

9) $\left(-\frac{24}{x}\right) - 10 + 5x$, $x = 4$

10) $(-4) + \frac{4x}{9}$, $x = 81$

Score: ....

Answer Key

1) 9	2) −10
3) 28	4) −4
5) 9	6) −40
7) 6	8) 3
9) 4	10) 32

Evaluating Two Variables

To solve an algebraic expression, substitute a number for each variable and perform the mathematical operations.

EXAMPLE:

Solve this expression. $-3x + 5y, x = 2, y = -1$

First substitute 2 for x, and -1 for y, then:
$$-3x + 5y = -3(2) + 5(-1) = -6 - 5 = -11$$

PRACTICES:

Simplify each algebraic expression.

1) $5a - (5 - b)$,
   $a = 2, b = 3$

2) $5x + 3y - 6 + 3y$,
   $x = 3, y = 1$

3) $\left(-\frac{27}{x}\right) + 4 + 3y$,
   $x = 3, y = 5$

4) $(-4)(-3a - 5b)$,
   $a = 3, b = 4$

5) $7x + 10 - 5y$,
   $x = 3, y = 6$

6) $18 + 3(-x - 4y)$,
   $x = 2, y = 5$

7) $12x + 2y$,
   $x = 5, y = 10$

8) $x \times 6 \div 3y$,
   $x = 6, y = 1$

9) $4x - 3y$,
   $x = 6, y = 3$

10) $\left(-\frac{14}{x}\right) + 4y$,
    $x = 7, y = -3$

Score: ....

Answer Key

1) 8	2) 15
3) 10	4) 116
5) 1	6) −48
7) 80	8) 12
9) 15	10) −14

Expressions and Variables

- In algebra, a variable is a letter used as a replacement for a number. The most common letters are: $x, y, z, a, b, c, m,$ and n.
- An algebraic expression is an expression that has variables, integers, and math operations such as addition, subtraction, multiplication, division, etc.
- In an expression, we can combine "identical" terms. (Values with same power and variable)

EXAMPLE:

Simplify this expression. $(10x + 2x + 3) = ?$

Combine like terms. Then: $(10x + 2x + 3) = 12x + 3$ (remember you cannot combine variables and numbers

PRACTICES:

Simplify each expression.

1) $10(-3 - 8x), x = 4$	2) $-3(5 - 8x) - 6x, x = 1$
3) $2x - 8x, x = 2$	4) $x + 12x, x = 6$
5) $20 - 5x + 10x + 5, x = 3$	6) $15(5x + 3), x = 0$
7) $20(4 - x) - 9, x = 2$	8) $20x - 8x - 10, x = 5$
9) $6x + 9y, x = 4, y = 2$	10) $6x - 2x, x = 8$

Score: ....

Answer Key	
1) -350	2) 3
3) -12	4) 78
5) 40	6) 45
7) 31	8) 50
9) 42	10) 32

Combining like Terms

- ✓ We separate the terms by "+" and "-" signs.
- ✓ Identical terms are those terms with same powers and same variables. Make sure to use the "+" or "-" that is in front of the coefficient

EXAMPLE:

Simplify this expression. $(-5)(8x - 6) =$

Use Distributive Property formula: $a(b + c) = a + ac$

$(-5)(8x - 6) = -40x + 30$

PRACTICES:

Simplify each expression.

1) $-8(-5x + 1)$

2) $6(-2 + 4x)$

3) $-8 - 14x + 16x + 3$

4) $9x - 7x - 15 + 18$

5) $(-9)(12x - 21) + 31$

6) $2(4x + 9) + 12x$

7) $4(-2x - 17) + 14(3x + 1)$

8) $(9x - 5y)7 + 25y$

9) $4.5x^3 \times (-8x)$

10) $-19 - 15x^2 + 12x^2$

Score: ....

Answer Key

1) $40x - 8$	2) $24x - 12$
3) $2x - 5$	4) $2x + 3$
5) $220 - 108x$	6) $20x + 18$
7) $34x - 54$	8) $63x - 10y$
9) $-36x^4$	10) $-3x^2 - 19$

Simplifying Polynomial Expressions

- A polynomial is a unique expression that consists of coefficients and variables that performs only the arithmetic of addition, subtraction, multiplication, and non-negative integer exponents of variables.

$$P(x) = a_n x^n + a_{n-1} x^{n-1} + \ldots + a_2 x^2 + a_1 x + a_0$$

EXAMPLE:

Simplify this Polynomial Expressions. $4x^2 - 5x^3 + 15x^4 - 12x^3 =$

Combine "like" terms: $-5x^3 - 12x^3 = -17x^3$

Then: $4x^2 - 5x^3 + 15x^4 - 12x^3 = 4x^2 - 17x^3 + 15x^4$

Then write in standard form: $4x^2 - 17x^3 + 15x^4 = 15x^4 - 17x^3 + 4x^2$

PRACTICES:

Simplify each polynomial.

1) $(2x^2 + 4) - (9 + 5x^2)$	2) $(25x^3 - 12x^2) - (6x^2 - 9x^3)$
3) $14x^5 - 15x^6 + 2x^5 - 16x^6 + x^6$	4) $(15 + 12x^3) + (3x^3 + 5)$
5) $13x^3 - 15x^4 + 12x^3 + 20x^4$	6) $-6x^2 + 15x^2 + 17x^3 + 16 - 32$
7) $15x^3 + 12 + 2x^2 - 5x - 10x$	8) $24x^2 - 16x^3 - 4x(2x^2 + 3x)$
9) $(21x^4 - 10x) - (2x - x^4)$	10) $(7x^2 - 9) + (x^2 - 8x^3)$

Score: ....

Answer Key

1) $-3x^2 - 5$	2) $34x^3 - 18x^2$
3) $-30x^6 + 16x^5$	4) $15x^3 + 20$
5) $5x^4 + 25x^3$	6) $17x^3 + 9x^2 - 16$
7) $15x^3 + 2x^2 - 15x + 12$	8) $-24x^3 + 12x^2$
9) $22x^4 - 12x$	10) $-8x^3 + 8x^2 - 9$

Chapter 6 : Equations and Inequalities

Topics that you'll learn in this chapter:

- One, Two, and Multi – Step Equations
- Graphing Single– Variable Inequalities
- One, Two, and Multi – Step Inequalities
- Solving Systems of Equations by Substitution and Elimination
- Finding Slope and Writing Linear Equations
- Graphing Lines Using Slope– Intercept and Standard Form
- Graphing Linear Inequalities
- Finding Midpoint and Distance of Two Points

"The study of mathematics, like the Nile, begins in minuteness but ends in magnificence." – *Charles Caleb Colton*

One-Step Equations

- ✓ The values of two algebraic expressions on both sides of an equation are always equal.
$$ax + b = c$$
- ✓ You only require performing one Math operation to solve the problem.

EXAMPLE:

Solve this equation. $x + 24 = 0$, $x = ?$

Here, we have the addition operation, and its inverse operation is subtraction. To solve this equation, subtract 24 from both sides of the equation: $x + 24 - 24 = 0 - 24$

Then simplify: $x + 24 - 24 = 0 - 24 \rightarrow x = -24$

PRACTICES:

Solve each equation.

1) $x + 4 = 16$	2) $48 = (-2) + x$
3) $5x = (-105)$	4) $(-8) = (8x)$
5) $(-2) = 14 + x$	6) $5 + x = 6$
7) $2x + 3 = (-7)$	8) $28 = x + 7$
9) $(-15) + x = (-15)$	10) $12x = (-36)$

Score: ....

Answer Key

1) 12	2) 50
3) −21	4) −1
5) −16	6) 1
7) −5	8) 21
9) 0	10) −3

Two-Step Equations

- You only require performing two math operations (add, subtract, multiply, or divide) to solve the equation.
- Simplify the equation using the inverse of addition or subtraction.
- Simplify the equation further by using the inverse of division or multiplication.

EXAMPLE:

Solve this equation. $3x = 15, x = ?$

Here, we have the multiplication operation (variable x is multiplied by 3) and its inverse operation is division. To solve this problem, we divide both sides of equation by 3:

$3x = 15 \rightarrow 3x \div 3 = 15 \div 3 \rightarrow x = 5$

PRACTICES:

Solve each equation.

1) $4(2 + 2x) = 8$

2) $(-5)(x - 3) = 25$

3) $(-5)(2x - 5) = (-15)$

4) $4(9 + 3x) = -12$

5) $6(2x + 1) = 30$

6) $2(x + 2) = 42$

7) $2(12 + 6x) = 60$

8) $(-10)(5x) = 100$

9) $4(3x + 3) = 24$

10) $\dfrac{x - 5}{3} = 4$

Score: ....

Answer Key	
1) 0	2) −2
3) 4	4) −4
5) 2	6) 10
7) 3	8) −2
9) 1	10) 17

Name: ....

Multi-Step Equations

✓ Combine "identical" terms on one side.

✓ Put variables to one side by adding or subtracting.

✓ Simplify the equation by using the inverse of addition or subtraction.

✓ Simplify further by using the inverse of division or multiplication.

EXAMPLE:

Solve this equation. $-(2 - x) = 5$

First, use Distributive Property: $-(2 - x) = -2 + x$

Now by adding 2 to both sides of the equation, we can solve it.

$-2 + x = 5 \rightarrow -2 + x + 2 = 5 + 2$
Now simplify: $-2 + x + 2 = 5 + 2 \rightarrow x = 7$

PRACTICES:

Solve each equation.

1) $8 - 2x = 28$	2) $-10 = -(x + 7)$
3) $2x - 17 = (-x) + 1$	4) $-2x = (-3x) - 8$
5) $5(14 + 2x) + 3x = -x$	6) $x - 11 = x - 5 + 2x$
7) $15 + 2x = (-25) - 2x + 3x$	8) $-3(x - 3x) = 40 - 4x$
9) $24 + 8x + x = (-x + 4)$	10) $-8(1 + 5x) = 152$

Score: ....

Answer Key

1) −10	2) 3
3) 6	4) −8
5) −5	6) −3
7) −40	8) 4
9) −2	10) −4

Graphing Single–Variable Inequalities

- ✓ Inequality is like equations and uses symbols for "less than" (<) and "greater than" (>).
- ✓ To solve inequalities, we need to separate the variable. (Like in equations)
- ✓ Find the value of the inequality on the number line to graph an inequality.
- ✓ For greater than or less than draw open circle on the value of the variable.
- ✓ Use filled circle if there is an equal sign too.
- ✓ Draw a line to the left or to the right for less or greater than.

EXAMPLE:

Draw a graph for $x > 2$

Since, the variable is greater than 2, then we need to find 2 and draw an open circle above it. Then, draw a line to the right.

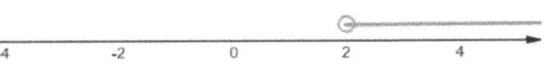

PRACTICES:

Draw a graph for each inequality.

1) $2 \geq x$

2) $x < 3$

3) $5 \geq x$

4) $x \geq -2$

5) $x > 0$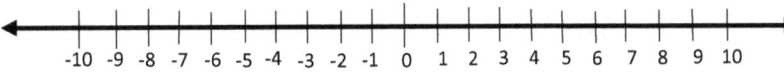

6) $-1.5 < x$

7) $x \geq -1$

Answer Key

1) $2 \geq x$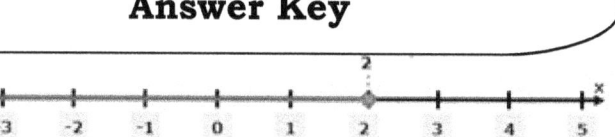

2) $x < 3$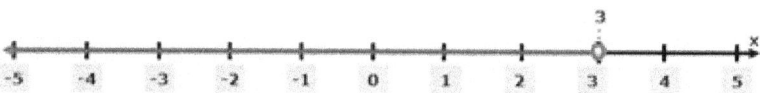

3) $5 \geq x$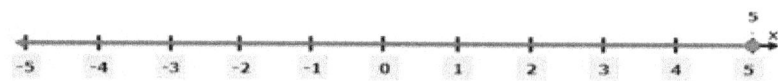

4) $x \geq -2$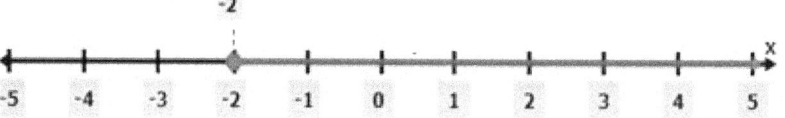

5) $x > 0$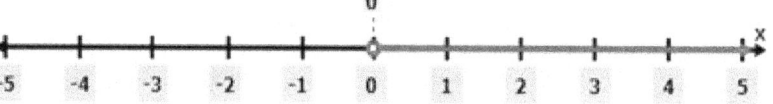

6) $-1.5 < x$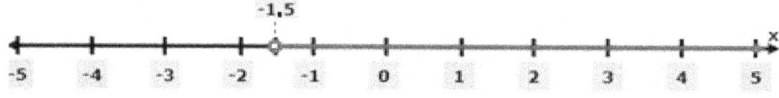

7) $x \geq -1$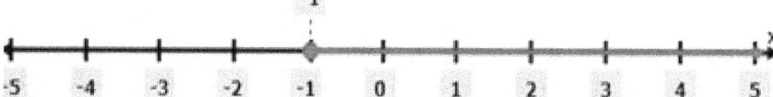

One–Step Inequalities

- ✓ Like equations, first separate the variable by using inverse operation.
- ✓ For dividing or multiplying both sides by negative numbers, flip the direction of the inequality sign.

EXAMPLE:

Solve this inequality. $x - 1 \leq 2$

Add 1 to both sides. $x - 1 \leq 2 \rightarrow x - 1 + 1 \leq 2 + 1$, then: $x \leq 3$

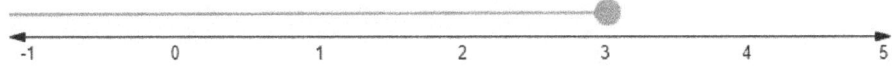

PRACTICES:

Solve each inequality and graph it.

1) $2x + 3 \geq 7$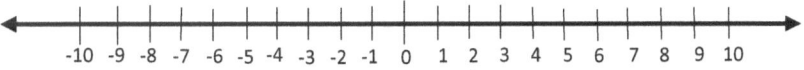

2) $x < 3$

3) $5 \geq x$

4) $x \geq -2$

5) $x > 0$

6) $-1.5 < x$

7) $x \geq -1$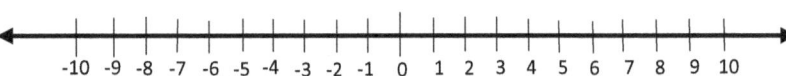

TASC Math Prep

Score: ....

Answer Key

1) $2 \geq x$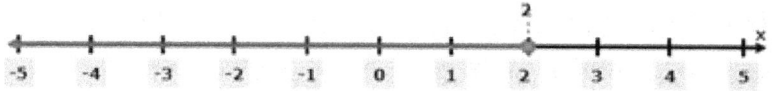

2) $x < 3$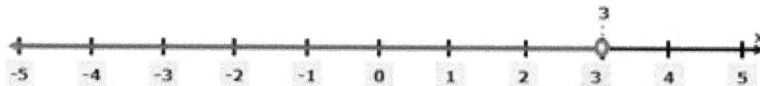

3) $5 \geq x$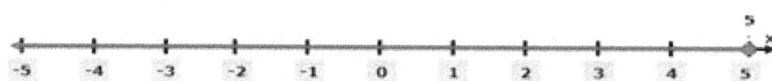

4) $x \geq -2$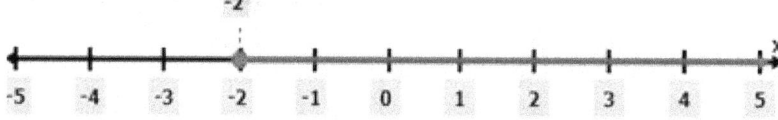

5) $x > 0$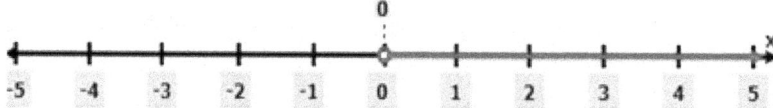

6) $-1.5 < x$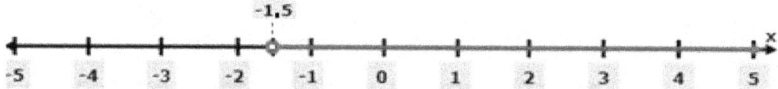

7) $x \geq -1$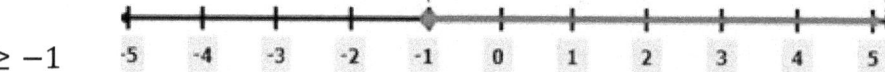

www.mathnotion.com

Two–Step Inequalities

- ✓ Separate the variable.
- ✓ Flip the direction of the inequality sign for dividing both sides by negatives numbers.
- ✓ We can simplify by using the inverse of addition or subtraction.
- ✓ We can simplify further by using the inverse of division or multiplication.

EXAMPLE:

Solve: $2x + 9 \geq 11$

First add -9 to both sides: $2x + 9 - 9 \geq 11 - 9 \rightarrow 2x \geq 2$

Now, divide both sides by 2: $2x \geq 2 \rightarrow x \geq 1$

PRACTICES:

Solve each inequality and graph it.

1) $x - 4 \leq 4$

2) $x + 4 \geq 5$

3) $3x - 2 \leq 7$

4) $5x + 2 < 12$

5) $x + 7 \geq 9$

6) $3x - 3 \leq 3$

7) $7x - 4 < 3$

8) $8 + x \leq 13$

9) $2x + 7 \leq 11$

10) $10x - 16 < 4$

Score: ....

Answer Key

1) $x \leq 8$	2) $x \geq 1$
3) $x \leq 3$	4) $x < 2$
5) $x \geq 2$	6) $x \leq 2$
7) $x < 1$	8) $x \leq 5$
9) $x \leq 2$	10) $x < 2$

Multi–Step Inequalities

- ✓ Separate the variable.
- ✓ We can simplify by using the inverse of addition or subtraction.
- ✓ We can simplify further by using the inverse of division or multiplication.

EXAMPLE:

Solve this inequality. $2x - 2 \leq 6$

First add 2 to both sides: $2x - 2 + 2 \leq 6 + 2 \rightarrow 2x \leq 8$

Now, divide both sides by 2: $2x \leq 8 \rightarrow x \leq 4$

PRACTICES:

Solve each inequality.

1) $-(x + 3) + 8 < 25$

2) $\frac{3x + 1}{2} \leq 5$

3) $\frac{x - 4}{3} > 7$

4) $4(x - 2) \leq 8$

5) $\frac{x}{3} + \frac{1}{3} < 2$

6) $\frac{x+4}{5} > 3$

7) $\frac{x}{8} + \frac{3}{4} < 1$

8) $2(x + 5) + 4 > 10$

9) $24 + 5x < 4$

10) $\frac{x+2}{3} > 10$

Score: ....

Answer Key

1) $x > -20$	2) $x \leq 3$
3) $x > 25$	4) $x \leq 4$
5) $x < 5$	6) $x > 11$
7) $x < 2$	8) $x > -2$
9) $x < -4$	10) $x > 28$

Solving Systems of Equations by Substitution

✓ Let the system of equations. $x + y = 1; -2x + y = 4$

Put $x = 1 - y$ in the second equation.

$-2(1 - y) + y = 4 \rightarrow -2 + 2y + y = 4 \Rightarrow y = 2$

Put $y = 2$ in $x = 1 - y$; then $x = 1 - 2 = -1$; $(-1, 2)$

EXAMPLE:

Solve: $-2x - 2y = -13; -4x + 2y = 10$

For the first equation above, you can add $-4x + 2y$ to the left side and 10 to the right side of the first equation: $-2x - 2y + (-4x + 2y) = -13 + 10$. Now, if you simplify, you get: $-2x - 2y - 4x + 2y = -3 \rightarrow -6x = -3 \rightarrow$

$x = 0.5$. Now, put 0.5 for the x in the first equation:

$-2(0.5) - 2y = -13$. By solving this equation, $y = 6$

PRACTICES:

Solve each system of equation by substitution.

1) $-x + 5y = -4$ $x - 3y = 8$	2) $2x + 3y = -6$ $-2x - y = 8$
3) $x + 2y = -5$ $5x - 10y = 5$	4) $y = -x + 5$ $3x - y = -3$
5) $3x = 6$ $10y = 4x + 2$	6) $3x + 2y = 2$ $x + 4y = -6$
7) $4x + y = 3$ $2x + 4y = -2$	8) $4y = 2x + 3$ $x - 4y = -2$
9) $7y = 14x$ $2x - 5y = -24$	10) $5y = x + 2$ $3x - 12y = -5$

Score: ....

Answer Key

1) $(14, 2)$	2) $(-\frac{9}{2}, 1)$
3) $(-2, -\frac{3}{2})$	4) $(\frac{1}{2}, \frac{9}{2})$
5) $(2, 1)$	6) $(2, -2)$
7) $(1, -1)$	8) $(-1, \frac{1}{4})$
9) $(3, 6)$	10) $(-\frac{1}{3}, \frac{1}{3})$

Name: ....

Solving Systems of Equations by Elimination

- ✓ A system of equations has two equations and two variables. For example, look at the system of equations: $x - y = 1, x + y = 5$
- ✓ The simplest way to solve a system of equation is using the elimination method. The elimination method uses the addition property of equality. On each side of an equation, you can add the same value.

EXAMPLE:

What is the value of x and y in this system of equations? $\begin{cases} 3x - 4y = -20 \\ -x + 2y = 10 \end{cases}$

Solving Systems of Equations by Elimination: $\begin{array}{l} 3x - 4y = -20 \\ -x + 2y = 10 \end{array}$ ⇒ Multiply the second equation by 3, then add it to the first equation.

$\begin{array}{l} 3x - 4y = -20 \\ 3(-x + 2y = 10) \end{array}$ ⇒ $\begin{array}{l} 3x - 4y = -20 \\ -3x + 6y = 30) \end{array}$ ⇒ $2y = 10 \Rightarrow y = 5$. Now, substitute 5 for y in the first equation and solve for x. $3x - 4(5) = -20 \rightarrow 3x - 20 = -20 \rightarrow x = 0$

PRACTICES:

Solve each system of equation by elimination.

1) $-5x + y = -5$ $-y = -6x + 6$	2) $-6x - 2y = -2$ $2x - 3y = 8$
3) $5x - 4y = 8$ $-6x + y = -21$	4) $10x - 4y = -24$ $-x - 20y = -18$
5) $25x + 3y = -13$ $12x - 6y = -36$	6) $x - 8y = -7$ $6x + 4y = 10$
7) $-6x + 16y = 4$ $5x + y = 11$	8) $2x - 3y = -10$ $4x + 6y = -20$
9) $x - 5y = -8$ $3x + 7y = -2$	10) $x - 2y = -3$ $2x + 6y = -1$

Answer Key

1) $(1, 0)$	2) $(1, -2)$
3) $(4, 3)$	4) $(-2, 1)$
5) $(-1, 4)$	6) $(1, 1)$
7) $(2, 1)$	8) $(-5, 0)$
9) $(-3, 1)$	10) $(-2, \frac{1}{2})$

Systems of Equations Word Problems

✓ Define your variables, write the system of equations then use elimination method for solving systems of equations.

EXAMPLE:

Tickets to a movie cost $5 for students and $8 for adults. Some friends purchased 20 tickets for $115.00. How many adults ticket did they buy?

Let x be the number of adult tickets and y be the number of student tickets. There are 20 tickets. Then: $x + y = 20$. The cost of students' ticket is $5 and for adults it is $8, and the total cost is $115. So, $8x + 5y = 20$. Now, we have a system of equations:

$$\begin{cases} x + y = 20 \\ 8x + 5y = 115 \end{cases}$$

Multiply the first equation by -5 and add to the second equation: $-5(x + y = 20) = -5x - 5y = -100$

$8x + 5y + (-5x - 5y) = 115 - 100 \to 3x = 15 \to x = 5 \to 5 + y = 20 \to y = 15$.

There are 5 adults' tickets and 15 student tickets.

PRACTICES:

Solve.

1) A school of 220 students went on a field trip. They took 20 vehicles, some vans, and some minibuses. Find the number of vans and the number of minibuses they took if each van holds 5 students and each minibus hold 15 students.

2) The sum of two numbers is 28. Their difference is 12. Find the numbers.

3) A farmhouse shelters 20 animals, some are pigs, and some are gooses. Altogether there are 64 legs. How many of each animal are there?

4) The sum of the digits of a certain two-digit number is 15. Reversing it's increasing the number by 9. What is the number?

TASC Math Prep

5) The difference of two numbers is 16. Their sum is 32. Find the numbers.

6) Tickets to a movie cost $5 for adults and $3 for students. A group of friends purchased 18 tickets for $82.00. How many adults ticket did they buy?

7) At a store, Eva bought two shirts and five hats for $154.00. Nicole bought three same shirts and four same hats for $168.00. What is the price of each shirt?

8) A farmhouse shelters 10 animals, some are pigs, and some are ducks. Altogether there are 36 legs. How many pigs are there?

9) A class of 195 students went on a field trip. They took 19 vehicles, some cars and some buses. If each car holds 5 students and each bus hold 25 students, how many buses did they take?

10) The sum of two numbers is 50. Their difference is 20. Find the numbers.

Score: ....

Answer Key

1) There are 8 van and 12 minibuses.	2) 8 and 20
3) There are 12 pigs and 8 gooses.	4) 78
5) 24 and 8.	6) 14
7) $32	8) 8
9) 5	10) 15 and 35

Linear Equations

- The equation of a line: $y = mx + b$
- Find the slope.
- Find the y-intercept. This can be done by substituting the slope and the coordinates of a point (x, y) on the line.

EXAMPLE:

What will be the equation of the line that passes through $(2, -2)$ and has a slope of 7?

$y = mx + b$ is the general slope-intercept form of the equation; where, m is the slope and b is the y-intercept.

By substitution of the given point and given slope, we have: $-2 = (2)(7) + b$

So, $b = -2 - 14 = -16$, and our required equation is $y = 7x - 16$.

PRACTICES:

Find the slope of the line through each pair of points.	Write the slope–intercept form of the equation of the line through the given points.
1) $(3, 1), (2, 4)$	2) Through: $(2, 3), (4, 2)$
3) $(-3, 4), (-1, 6)$	4) Through: $(8, -3), (6, 7)$
5) $(4, 4), (6, -6)$	6) Through: $(0.5, 4), (2.5, 4.4)$
7) $(-1, 8), (5, -4)$	8) Through: $(4, -2), (2.5, 1)$
9) $(12, -3), (7, -3)$	10) Through: $(-1, 0.7), (-2.3, 2)$

Score: ....

Answer Key

1) -3	2) $y = -\frac{1}{2}x + 4$
3) 1	4) $y = -5x + 37$
5) -5	6) $y = \frac{1}{5}x + \frac{39}{10}$
7) -2	8) $y = -2x + 6$
9) 0	10) $y = -x - 0.3$

Graphing Lines of Equations

✓ The equation of the line is: $y = mx + c$

If slope-intercept form of a line given the slope m and the y-intercept (the intersection of the line and y-axis)

EXAMPLE:

Sketch the graph of $y = 8x - 3$.

To graph this line, two points are required.

We have value of y is -3 when x is 0.

The value of x is 3/8 when y is 0.

$x = 0 \rightarrow y = 8(0) - 3 = -3$,

$y = 0 \rightarrow 0 = 8x - 3 \rightarrow x = \frac{3}{8}$

Now, the two points are: $(0,-3)$ and $(\frac{3}{8},0)$.

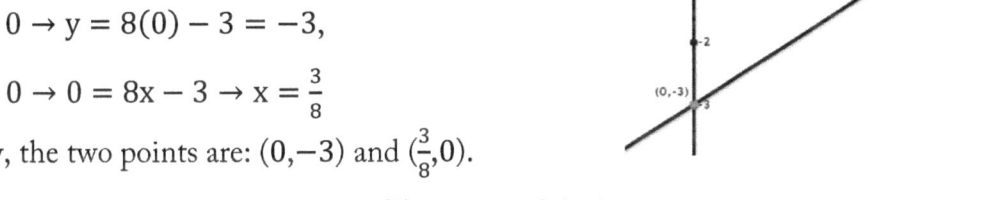

Find the points and graph the line. The slope of the line is 8.

PRACTICES:

Sketch the graph of each line.

1) $y = 3x - 2$

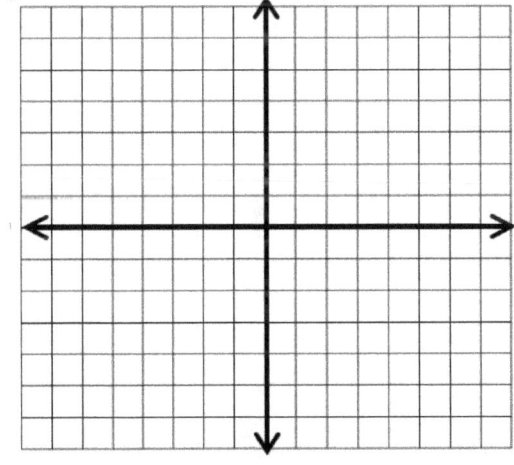

2) $y = 2x + 3$

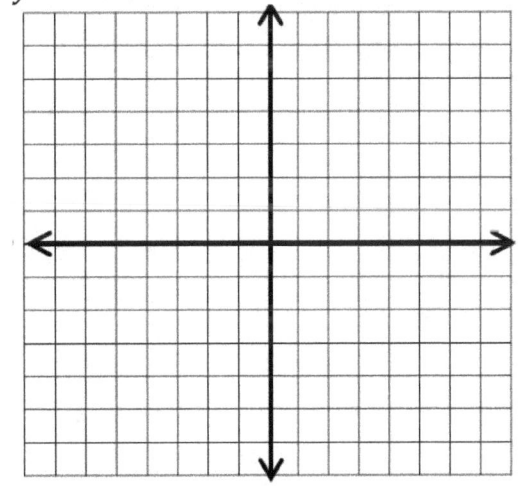

3) $-2x = y + 5$

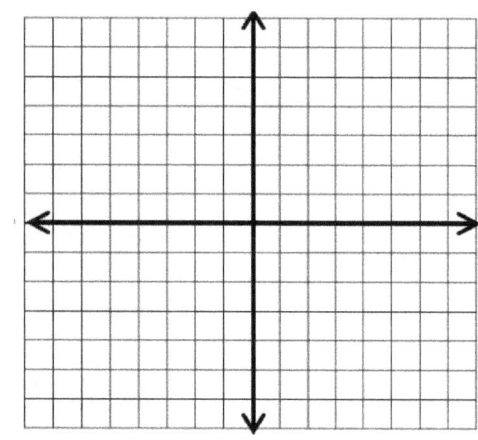

4) $4x + y = 2$

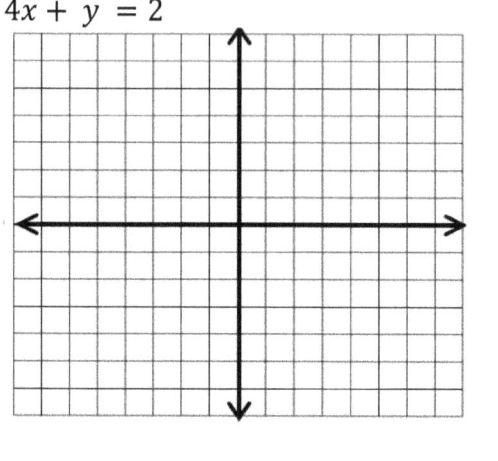

Score: ....

Answer Key

1)
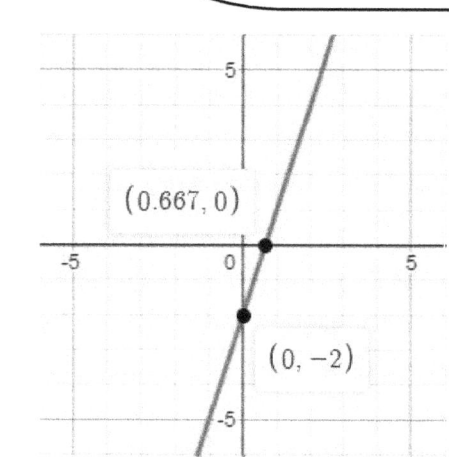
$(0.667, 0)$
$(0, -2)$

2)
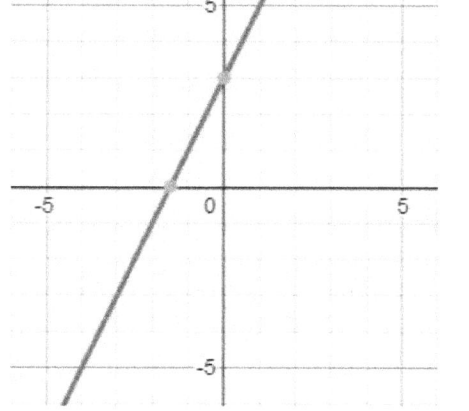

3)
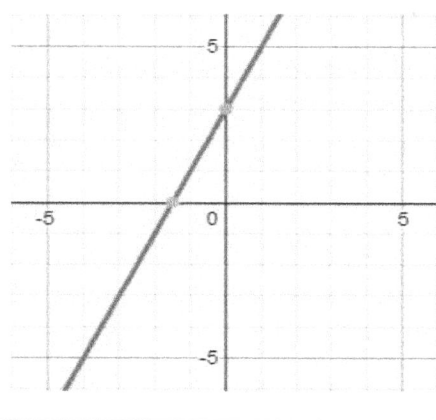

4)
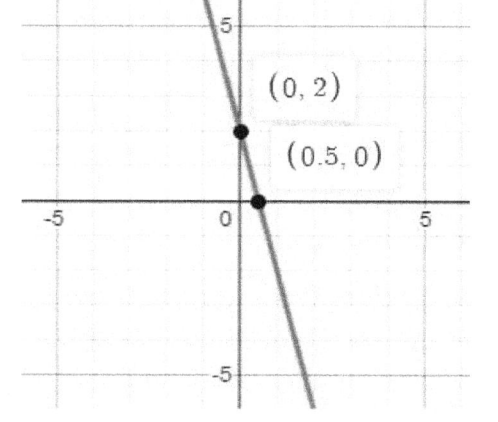
$(0, 2)$
$(0.5, 0)$

Graphing Linear Inequalities

- ✓ First step is to graph the "equals" line.
- ✓ Choose a testing point. (It can be any point on both sides of the line.)
- ✓ Put the value of (x, y) of that point in the inequality. If this satisfy the inequality, then this part of line is solution. If it does not satisfy then other part of line, is solution.

EXAMPLE:

Plot the graph of $y < 2x - 3$.

We know that first step is to graph the line: $y = 2x - 3$.

Now y-intercept is -3 and slope is 2. Then, select a testing point. The simplest point to test is the origin: $(0, 0)$

$(0,0) \to y < 2x - 3 \to 0 < 2(0) - 3 \to 0 < -3$

0 is greater than -3. So, the other part of the line (On the right side) is the solution.

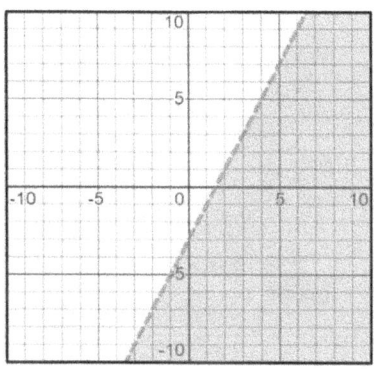

PRACTICES:

Sketch the graph of each linear inequality.

1) $2y + 8x \geq 4$

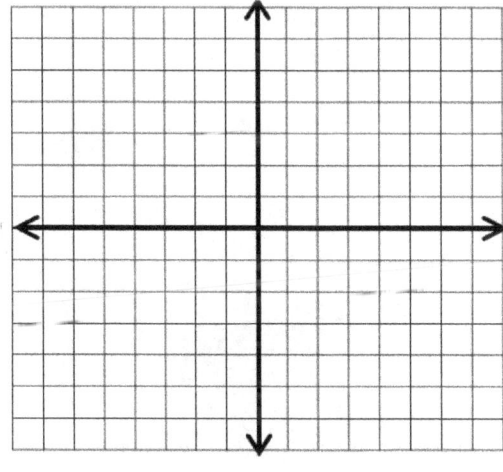

2) $-x + 2y \leq 4$

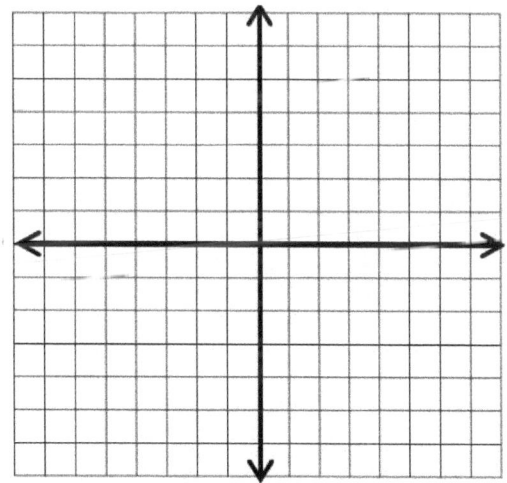

3) $2x + \frac{1}{2}y < 2$

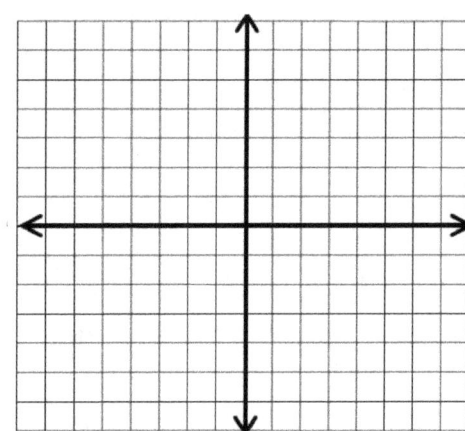

4) $-\frac{1}{3}x + y < 2$

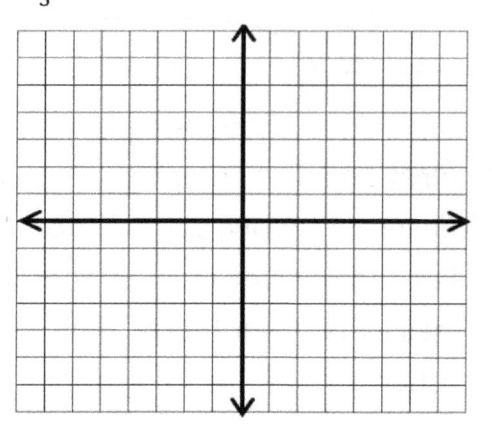

Score: ....

Answer Key

1)
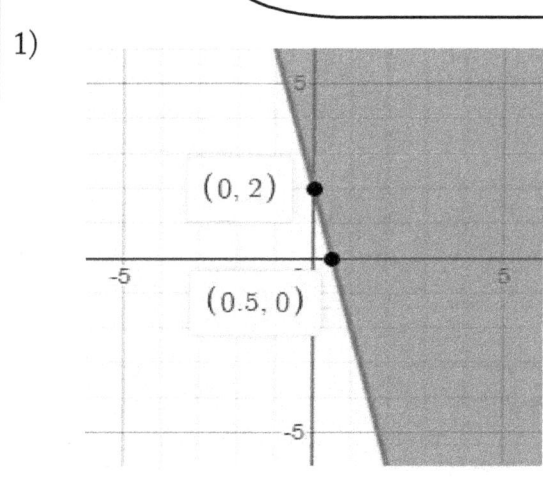
(0, 2)
(0.5, 0)

2)
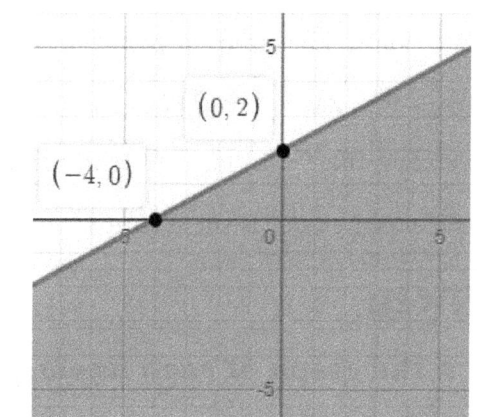
(0, 2)
(−4, 0)

3)
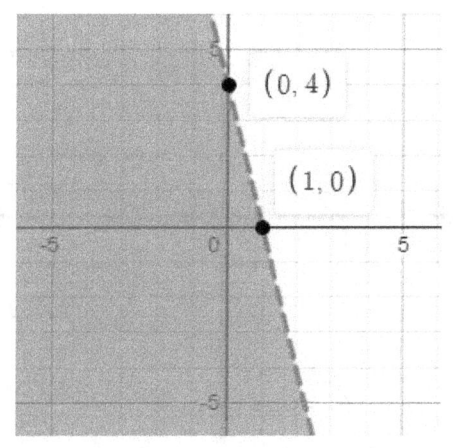
(0, 4)
(1, 0)

4)
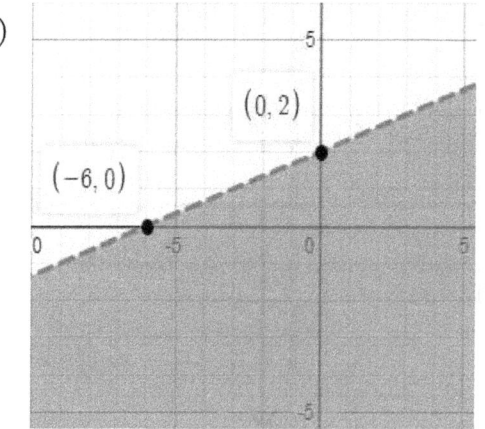
(0, 2)
(−6, 0)

Finding Distance of Two Points

- Distance of wo points A (x_1, y_1) and B (x_2, y_2): $d = \sqrt{(x_1 - x_2)^2 + (y_1 - y_2)^2}$
- The midpoint is middle of a line segment.
- We can find Midpoint of two endpoints A (x_1, y_1) and B (x_2, y_2) using this formula: $M\left(\frac{x_1+x_2}{2}, \frac{y_1+y_2}{2}\right)$

EXAMPLE:

Find the distance between of $(0, 8), (-4, 5)$.

Use distance of two points formula: $d = \sqrt{(x_1 - x_2)^2 + (y_1 - y_2)^2}$

$(x_1, y_1) = (0, 8)$ and $(x_2, y_2) = (-4, 5)$. Then: $d = \sqrt{(x_1 - x_2)^2 + (y_1 - y_2)^2} \to$

$d = \sqrt{(0 - (-4))^2 + (8 - 5)^2} = \sqrt{(4)^2 + (3)^2} = \sqrt{16 + 9} = \sqrt{25} = 5 \to d = 5$

PRACTICES:

Find the midpoint of the line segment with the given endpoints.	Find the distance between each pair of points.
1) $(1.5, -1), (0.5, -1)$	2) $(3, 4), (2, -1)$
3) $(1.5, -1), (0.5, -1)$	4) $(6, -1), (2, 3)$
5) $(0, 3), (4, -9)$	6) $(2, 5), (-2, 5)$
7) $(5, 2), (1, 5)$	8) $(0, -4), (-5, 1)$
9) $(-2, 0), (3, -4)$	10) $(3, -2), (-1, -5)$

Answer Key

1) $(1, -1)$	2) 5.09
3) $(1.5, -2)$	4) 5.656
5) $(2, -3)$	6) 4
7) $(3, 3.5)$	8) 7.07
9) $(0.5, -2)$	10) 5

Chapter 7 : Polynomials

Topics that you'll learn in this chapter:

- Classifying Polynomials
- Writing Polynomials in Standard Form
- Simplifying Polynomials
- Adding and Subtracting Polynomials
- Multiplying and Dividing Monomials
- Multiplying a Polynomial and a Monomial
- Multiplying Binomials
- Factoring Trinomials
- Operations with Polynomials

"Mathematics – the unshaken Foundation of Sciences, and the plentiful Fountain of Advantage to human affairs." — Isaac Barrow

Classifying Polynomials

Name	Degree	Example
constant	0	4
linear	1	$2x$
quadratic	2	$x^2 + 5x + 6$
cubic	3	$x^3 - x^2 + 4x + 8$
quartic	4	$x^4 + 3x^3 - x^2 + 2x + 6$
quantic	5	$x^5 - 2x^4 + x^3 - x^2 + x + 10$

EXAMPLE:

$17x^5 \Rightarrow$ Quantic binomial

PRACTICES:

Name each polynomial by degree and number of terms.	Write each polynomial in standard form
1) -5	2) $12x^4 + x - 4x^3$
3) $x + 1$	4) $12 - x^3 - 3x^5 + 9x^4$
5) $8x^6 - 7$	6) $x^2 + 13x^5 + x^3 - 4x$
7) $3x^2 - x$	8) $x^5 + 2x^3 (x^2 + 2)$
9) $-8x^4 + 3x^3 - 2x^2 - 3x$	10) $(x - 5)(x + 5)$

Answer Key

1) Constant monomial	2) $12x^4 - 4x^3 + x$
3) Linear binomial	4) $-3x^5 + 9x^4 - x^3 + 12$
5) Sixth degree binomial	6) $13x^5 + x^3 + x^2 - 4x$
7) Quadratic binomial	8) $2x^6 + x^5 + 4x^3$
9) Quartic polynomial with four terms	10) $x^2 - 25$

Name: ....

Adding and Subtracting Polynomials

- ✓ To add polynomials, we combine like terms with some order of operations considerations thrown in.
- ✓ You have to be careful with a negative sign and don't confuse it with addition and subtraction.

EXAMPLE:

Add expressions. $(2x^3 - 6) + (9x^3 - 4x^2) = ?$

Remove parentheses: $(2x^3 - 6) + (9x^3 - 4x^2) = 2x^3 - 6 + 9x^3 - 4x^2$

Now combine like terms: $2x^3 - 6 + 9x^3 - 4x^2 = 11x^3 - 4x^2 - 6$

PRACTICES:

Simplify each expression.

1) $(x^3 + 6) - (6 + 3x^3)$

2) $(x^2 + 8) + (7x^2 - 8)$

3) $(2x^2 + x^3) - (5x^2 + 1)$

4) $(6x^2 - 4x) + (3x - 6x^2 + 1)$

5) $(x - 2x^3) - (4x^3 + 4)$

6) $(2x^3 + 2x^2) - (2x^2 - x^3 + 2)$

7) $(4x^2 - 3) + (x^2 - x^3)$

8) $(x^3 + 13x^4) - (13x^4 + 3x^3)$

9) $(x^4 + 2x^5 + 3x^3) + (4x^3 + 6x^4)$

10) $(3x^3 - 6x^6) + (3x^3 + 2x^6)$

Score: ....

Answer Key

1) $-2x^3$	2) $8x^2$
3) $x^3 - 3x^2 + 1$	4) $-x + 1$
5) $-6x^3 + x - 4$	6) $3x^3 - 2$
7) $-x^3 + 5x^2 - 3$	8) $-2x^3$
9) $2x^5 + 7x^4 + 7x^3$	10) $-4x^6 + 6x^3$

Multiply and Divide Monomials

- When dividing monomials, we divide coefficients first and then divide their variables.
- If we have exponents with the same base, we'll subtract their powers.
- Exponent's rules:

$$x^a \times x^b = x^{a+b}, \quad \frac{x^a}{x^b} = x^{a-b}$$

$$\frac{1}{x^b} = x^{-b}, \quad (x^a)^b = x^{a \times b}$$

$$(xy)^a = x^a \times y^a$$

EXAMPLE:

Multiply expressions. $(-3x^7)(4x^3) = ?$

Use this formula: $x^a \times x^b = x^{a+b} \rightarrow x^7 \times x^3 = x^{10}$; Then: $(-3x^7)(4x^3) = -12x^{10}$

Dividing expressions. $\frac{18x^2y^5}{2xy^4} = ?$

Use this formula: $\frac{x^a}{x^b} = x^{a-b}$, $\frac{x^2}{x} = x^{2-1} = x$ and $\frac{y^5}{y^4} = y^{5-4} = y$; Then: $\frac{18x^2y^5}{2xy^4} = 9xy$

PRACTICES:

Simplify.

1) $(x^3y^2)(42y^4)$

2) $\frac{100x^5y^6}{25x^6y^{11}}$

3) $(8x^4)(12x^5)$

4) $\frac{75x^{16}y^{10}}{5x^6y^7}$

5) $(-2x^{-3}y^2)^2$

6) $\frac{15x^{12}y^5}{5x^9y^2}$

7) $(11x^2y^4)(4x^9y^{10})$

8) $\frac{50x^4y^7}{25x^3y^7}$

9) $(2x^{-3}y^4)^2$

10) $\frac{-21x^8y^{13}}{3x^6y^6}$

Answer Key

1) $42x^3y^6$	2) $4x^{-1}y^{-5}$
3) $96x^9$	4) $15x^{10}y^3$
5) $4x^{-6}y^4$	6) $3x^3y^3$
7) $44x^{11}y^{14}$	8) $2x$
9) $4x^{-6}y^8$	10) $-7x^2y^7$

Name: ....

Multiplying Monomials

- A polynomial having only one term is called monomial, like $2x$ or $7y$

EXAMPLE:

Multiply expressions. $5a^4b^3 \times 2a^3b^2 = ?$

Use this formula: $x^a \times x^b = x^{a+b}$

$a^4 \times a^3 = a^{4+3} = a^7$ and $b^3 \times b^2 = b^{3+2} = b^5$

Then: $5a^4b^3 \times 2a^3b^2 = 10a^7b^5$

PRACTICES:

Simplify each expression.

1) $2xy^2 \times 3z^2$

2) $3xyz \times 5x^2y$

3) $4pq^3 \times (-3p^3q)$

4) $s^3t^2 \times 2st^5$

5) $5p^3 \times (-2p^2)$

6) $-2p^2r \times 6pr^3$

7) $(-a)(-4a^6b)$

8) $2u^2v^3 \times (-8u^3v^3)$

9) $6u^3 \times (2u)$

10) $-5y^2 \times 4x^2y$

Score: ....

Answer Key

1) $6xy^2z^2$	2) $15x^3y^2z$
3) $-12p^4q^4$	4) $2s^4t^{10}$
5) $-10p^5$	6) $-12p^3r^4$
7) $4a^7b$	8) $-16u^5v^6$
9) $12u^4$	10) $-20x^2y^3$

Multiply a Polynomial and a Monomial

- Use the product rule for exponents to multiply monomials.
- Use distributive property to multiply a monomial by a polynomial.

$$a \times (b + c) = a \times b + a \times c$$

EXAMPLE:

Multiply expressions. $-4x(5x + 9) = ?$

Use Distributive Property: $-4x(5x + 9) = -20x^2 - 36x$

PRACTICES:

Find each product.

1) $3(2x - 2y)$

2) $5x(4x - y)$

3) $-2x(x + 5)$

4) $11(3x + 7)$

5) $10x(5x - 2y)$

6) $4(3x - 5y)$

7) $2x(3x^3 - 5x + 4)$

8) $-4x(2 + 4xy)$

9) $3(2x^2 - 8x + 3)$

10) $-3x^2(3x^2 + 5)$

Score: ....

Answer Key

1) $6x - 6y$	2) $20x^2 - 5xy$
3) $-2x^2 - 10$	4) $33x + 77$
5) $50x^2 - 20xy$	6) $50x^2 - 20xy$
7) $6x^4 - 10x^2 + 8x$	8) $-16x^2y - 8x$
9) $6x^2 - 24x + 9$	10) $-9x^4 - 15x^2$

Multiply Binomials

- Use "FOIL". (First-Out-In-Last)

$$(x + a)(x + b) = x^2 + (b + a)x + ab$$

EXAMPLE:

Multiply Binomials. $(x + 5)(x - 2) = ?$

Use "FOIL". (First–Out–In–Last):

$(x + 5)(x - 2) = x^2 - 2x + 5x - 10$
Then simplify: $x^2 - 2x + 5x - 10 = x^2 + 3x - 10$

PRACTICES:

Multiply.

1) $(2x - 2)(x + 3)$

2) $(4x + 2)(2x + 1)$

3) $(x + 3)(x + 4)$

4) $(x^2 + 5)(x^2 - 5)$

5) $(2x - 3)(x + 4)$

6) $(2x - 6)(x + 7)$

7) $(x - 2)(3x - 4)$

8) $(2x - 5)(x + 4)$

9) $(x + 10)(x - 10)$

10) $(x - 3)(3x + 4)$

Score: ....

Answer Key

1) $2x^2 + 4x - 6$	2) $8x^2 + 8x + 2$
3) $x^2 + 7x + 12$	4) $x^4 - 25$
5) $2x^2 + 5x - 12$	6) $2x^2 + 8x - 42$
7) $3x^2 - 10x + 8$	8) $2x^2 + 3x - 20$
9) $x^2 - 100$	10) $3x^2 - 5x - 12$

Name: ....

Factor Trinomials

- ✓ FOIL:
$$(x+a)(x+b) = x^2 + (b+a)x + ab$$
- ✓ "Difference of Squares":
$$a^2 - b^2 = (a+b)(a-b)$$
$$a^2 + 2ab + b^2 = (a+b)(a+b)$$
$$a^2 - 2ab + b^2 = (a-b)(a-b)$$
- ✓ "Reverse FOIL":
$$x^2 + (b+a)x + ab = (x+a)(x+b)$$

EXAMPLE:

Factor this trinomial. $x^2 - 2x - 8 = ?$

Expression is broken into groups: $(x^2 + 2x) + (-4x - 8)$

Now x is a factor from $x^2 + 2x$: $x(x+2)$ and factor out -4 from $-4x - 8$: $-4(x+2)$

Then: $= x(x+2) - 4(x+2)$, now factor out like term: $x + 2$

Then: $(x+2)(x-4)$

PRACTICES:

Factor each trinomial.

1) $x^2 - 12x + 27$	2) $x^2 + 5x - 24$
3) $x^2 + 13x + 30$	4) $x^2 - 81$
5) $2x^2 + 12x - 14$	6) $x^2 + 2x - 8$
7) $2x^2 + 3x + 1$	8) $2x^2 + 2x - 4$
9) $9x^2 + 3x - 2$	10) $x^2 + 15x + 56$

Score: ....

Answer Key

1) $(x-3)(x-9)$	2) $(x+8)(x-3)$
3) $(x+10)(x+3)$	4) $(x+9)(x-9)$
5) $(x+7)(2x-2)$	6) $(x-2)(x+4)$
7) $(2x+1)(x+1)$	8) $(2x-2)(x+2)$
9) $(3x-1)(3x+2)$	10) $(x+7)(x+8)$

Operations with Polynomials

- Use distributive property to multiply a monomial by a polynomial.

$$a \times (b + c) = a \times b + a \times c$$

EXAMPLE:

Multiply. $5(2x - 6) =$

Use the distributive property: $5(2x - 6) = 5 \times 2x - 5 \times (-6) = 10x - 30$

PRACTICES:

Find each product.

1) $x^2(3x - 2)$

2) $2x^2(5x - 3)$

3) $-x(5x - 3)$

4) $x^2(-3x + 9)$

5) $5(7x + 3)$

6) $8(3x + 8)$

7) $5(10x + 4)$

8) $-3x^5(x - 3)$

9) $5(3x^2 - x + 2)$

10) $4(x^2 - 2x + 3)$

Score: ....

Answer Key

1) $3x^3 - 2x^2$	2) $10x^3 - 6x^2$
3) $-5x^2 + 3x$	4) $-3x^3 + 9x^2$
5) $35x + 15$	6) $24x + 64$
7) $50x + 20$	8) $-3x^6 + 9x^5$
9) $15x^2 - 5x + 10$	10) $4x^2 - 8x + 12$

Simplifying Polynomials

- ✓ Find "identical" terms. (They have same variables with same power).
- ✓ Use "FOIL". (First-Out-In-Last) for binomials:
$$(x + a)(x + b) = x^2 + (b + a)x + ab$$
- ✓ Using order of operation, add or subtract "identical" terms

EXAMPLE:

Simplify this expression. $(4 + x)(x - 3) = ?$

Use FOIL: $(x + 4)(x - 3) = x^2 + x - 12$

PRACTICES:

Simplify each expression.

1) $-3x^2 + x^5 + 7x^5 - 2x^2 + 6$	2) $18x^5 - 3x^5 + 7x^2 - 15x^5 + 4$
3) $x(x^3 + 9) - 6(8 + x^2)$	4) $x(x^2 + 2x^3) - x^3 + x$
5) $4 - 17x^2 + 30x^2 - 17x^2 + 26$	6) $4x^2 - 8x + 3x^3 + 15x - 20x$
7) $(x - 6)(x - x^2 + 5)$	8) $(x - 5)(x + 5)$
9) $(x^4 - x) + (4x^2 - 3x^4)$	10) $x(x^2 + x + 3)$

Score: ....

Answer Key

1) $8x^5 - 5x^2 + 6$	2) $7x^2 + 4$
3) $x^4 - 6x^2 + 9x - 48$	4) $2x^4 + x$
5) $-4x^2 + 30$	6) $3x^3 + 4x^2 - 13x$
7) $-x^3 + 7x^2 - x - 30$	8) $x^2 - 25$
9) $-2x^4 + 4x^2 - x$	10) $x^3 + x^2 + 3x$

Chapter 8 : Functions

Topics that you'll learn in this chapter:

- Relations and Functions
- Rate of change and Slope
- x and y intercept
- Slope-intercept form
- Slope-point form
- Equation of Parallel or Perpendicular lines
- Equation of Horizontal and Vertical Lines
- Function Notation
- Adding and Subtracting Functions
- Multiplying and Dividing Functions
- Composition of Functions
- Solve a Quadratic Equation

It's fine to work on any problem, so long as it generates interesting mathematics along the way – even if you don't solve it at the end of the day." – Andrew Wiles

Relations and Functions

- ✓ **RELATION:** A relationship between two sets of elements like input and output, input can have as many as outputs.
- ✓ **FUNCTION:** A relationship between two sets of elements like input and output and only and exactly one output related to one input.
- ✓ A function is a type of relation, but a relation may not be a function.
- ✓ **The Vertical Line Test**
  A graphical method that we can observe the cross between the graph and vertical line.
  The graph is a function if and only if the intersection point is one.

EXAMPLE:

Use the vertical line test to determine if the graph is function or not?

The graph is not a function

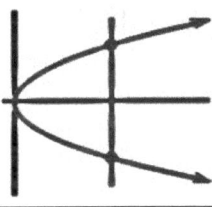

PRACTICES:

State the domain and range of each relation. Then determine whether each relation is a function.

1)
Function:
....
Domain:
....
Range:
....

2)
Function:
....
Domain:
....
Range:
....

3) $\{(1,-2),(4,-1),(0,5),(4,0),(3,8)\}$

Function:
....
Domain:
....
Range:
....

4)

Function:
....
Domain:
....
Range:
....

x	y
3	4
0	1
−2	−3
6	−3
8	2

Score: ....

Answer Key

1) No, $D_f = \{1, 3, 5, 7, 9\}$,
   $R_f = \{3, 5, 8, 12, 18\}$

2) Yes, $D_f = (-\infty, \infty)$,
   $R_f = \{2, -\infty)$

3) No, $D_f = \{1, 4, 0, 3\}$,
   $R_f = \{-2, -1, 5, 0, 8\}$

4) Yes, $D_f = \{3, 0, -2, 6, 8\}$,
   $R_f = \{4, 1, -3, 2\}$

Rate of change

- ✓ Slope can be described as "rate of change".
- ✓ Rate of change is a ratio between a change in one variable comparing to a corresponding change in another variable. Rate of change $= \frac{change\ in\ output\ (y)}{change\ in\ input\ (x)}$
- ✓ Rates of change can be positive, negative, or zero.

EXAMPLE:

The table shows the amount of money SB carwash made washing car. Find the rate of change in dollar per car?

SB Carwash	Number	4	8	12	16
	Money ($)	32	56	80	104

Rate of change $= \frac{change\ in\ output\ (y)}{change\ in\ input\ (x)} = \frac{Change\ in\ money}{Change\ in\ car} = \frac{56-32}{8-4} = \frac{24}{4} = \frac{6}{1}$, or $6 per car

PRACTICES:

What is the average rate of change of the function?

1)
Gallons	3	5	7	9
Miles	81	135	189	243

2)
Products	145	159	173	187
Costs	761	719	677	635

3)
x	4.5	6	7.5	9
y	6	15	24	33

4)
x	41	47	53	59
y	67	52	37	22

5) $f(x) = -2x + 4$, from $x = -1$ to $x = 4$?

6) $f(x) = x - 6$, from $x = -5$ to $x = 1$?

7) $f(x) = -4$, from $x = 3$ to $x = -2$?

8) $f(x) = 3x^2 + 5$, from $x = 3$ to $x = 6$?

9) $f(x) = -2x^2 - 4$, from $x = 2$ to $x = 4$?

10) $f(x) = x^3 + 3$, from $x = 1$ to $x = 2$?

Answer Key

1) 27 miles per gallon	2) −3 cost per product
3) 6	4) −2.5
5) −2	6) 1
7) 0	8) 27
9) −12	10) 7

Slope

- The slope is used to describe the steepness and direction of lines on the coordinate plane.
- A coordinate plane is a two-dimensional plane formed by the intersection of a vertical line called y-axis and a horizontal line called x-axis. These are perpendicular lines that intersect each other at zero, and this point is called the origin.
- An ordered pair (x, y) shows the location of a point.
- A line on coordinate plane can be drawn by connecting two points.
- The slope of a line with two points A (x_1, y_1) and B (x_2, y_2) can be found by using this formula: $\frac{y_2 - y_1}{x_2 - x_1} = \frac{rise}{run}$

EXAMPLE:

Use the given points to determine the slope. Points, $(3, -7)$, $(-2, -9)$

Slope = $\frac{y_2 - y_1}{x_2 - x_1} = \frac{-9 - (-7)}{-2 + 3} = \frac{-9 + 7}{1} = \frac{-2}{1} = -2$

PRACTICES:

Find the slope of the line through each pair of points.

1) $(2, -9), (5, -6)$	2) $(14, 7), (20, 12)$
3) $(1, -5), (8, -4)$	4) $(13, -9), (15, -7)$
5) $(-5, -8), (-8, -2)$	6) $(0, 0), (12, -2)$
7) $(14, -8), (-6, 5)$	8) $(-2, 5), (-2, 8)$
9) $(-14, -9), (-6, -15)$	10) $(-19, 2), (2, -19)$

Score: ....

Answer Key

1) 1	2) $\frac{5}{6}$
3) $\frac{1}{7}$	4) 1
5) -2	6) $-\frac{1}{6}$
7) $-\frac{13}{20}$	8) Undefined
9) $-\frac{3}{4}$	10) -1

x and y intercept

- ✓ x-intercept is the point at which the graph crosses the x-axis, and the value of y is zero.
- ✓ y-intercept is the point at which the graph crosses the y-axis, and the value of x is zero.

EXAMPLE:

Find the intercepts of the equation y = 3x − 21.

To find the x-intercept, set y = 0, then 0 = 3x − 21 → 3x = 21 → x = 7. x-intercept(7,0)

To find the y-intercept, set y = 0, then y = 3(0) − 21 → y = −21. y-intercept (0, −12)

PRACTICES:

Find the x and y intercepts for the following equations.

1) $5x + 3y = 15$	2) $y = x + 8$
3) $4x = y + 16$	4) $x + y = -2$
5) $4x - 3y = 7$	6) $7y - 5x + 10 = 0$
7) $\frac{3}{7}x + \frac{1}{4}y + \frac{2}{3} = 0$	8) $3x - 21 = 0$
9) $24 - 4y = 0$	10) $-2x - 6y + 42 = 12$

Answer Key

1) $y-intercept = 5$ $x-intercept = 3$	2) $y-intercept = 8$ $x-intercept = -8$
3) $y-intercept = -16$ $x-intercept = 4$	4) $y-intercept = -2$ $x-intercept = -2$
5) $y-intercept = -\frac{7}{3}$ $x-intercept = \frac{7}{4}$	6) $y-intercept = -\frac{10}{7}$ $x-intercept = 2$
7) $y-intercept = -\frac{8}{3}$ $x-intercept = -\frac{2}{7}$	8) $y-intercept = undefind$ $x-intercept = 7$
9) $y-intercept = 6$ $x-intercept = undefind$	10) $y-intercept = 5$ $x-intercept = 15$

Writing Linear Equations

- The equation of a line:
$$y = mx + b$$
- Identify the slope.
- Find the y-intercept. This can be done by substituting the slope and the coordinates of a point (x, y) on the line.

EXAMPLE:

If f is a linear function, with $f(4) = -2$, and $f(7) = 1$, find an equation for the function.

Slope: $\frac{y_2 - y_1}{x_2 - x_1} = \frac{1-(-2)}{7-4} = \frac{1+2}{3} = 1$

$y = mx + b \rightarrow y = 1 \times x + b \rightarrow y = x + b$, and $f(7) = 1$, then $1 = 7 + b \rightarrow b = -6$

The linear equation is: $y = x - 6$

PRACTICES:

Write the equation of each line in form of $y = mx + b$.

1) $m = 2$; y-intercept= -7	2) $m = -\frac{2}{5}$; y-intercept= $\frac{2}{3}$
3) $f(-2) = 1$; $f(-3) = 4$	4) $f(6) = -1$; $f(2) = 7$
5) Through: $(5, 7), (3, 6)$	6) Through: $(-1.5, 2), (6.5, -2)$
7) Through: $(2, -1), (6, 11)$	8) Through: $(4, 1), (-2, 7)$
9) Through: $(2, 4), (-2, -4)$	10) Through: $(4, 5), (0, -1)$

Answer Key

1) $y = 2x - 7$	2) $y = -\frac{2}{5}x + \frac{2}{3}$
3) $y = -3x - 5$	4) $y = -2x + 11$
5) $y = \frac{1}{2}x + \frac{9}{2}$	6) $y = -0.5x + 1.25$
7) $y = 3x - 7$	8) $y = -x + 5$
9) $y = 2x$	10) $y = 1.5x - 1$

Slope-intercept form

- ✓ The slope-intercept form is one of several ways you can write the equation of a line.
- ✓ Using the slope m and the y-intercept b, then the equation of the line is:
$$y = mx + b$$

EXAMPLE:

Solve for y when 4x - 2y = 12.

Subtract $4x$ from both sides: $-4x + 4x - 2y = 12 - 4x \rightarrow -2y = 12 - 4x$

Divide everything by -2: $y = -6 + 2x \rightarrow y = 2x - 6$

PRACTICES:

Write the slope–intercept form of the equation of each line.

1) $y - 4 = x + 3$	2) $5x + 14 = -3y$
3) $18x - 12y = -6$	4) $7x - 4y + 25 = 0$
5) $-\frac{1}{3}y = -2x + 3$	6) $5 - y - 4x = 0$
7) $-y = -6x - 9$	8) $-2(7x + y) = 24$
9) $3(y + 3) = 2(x - 3)$	10) $\frac{3}{4}y + \frac{1}{4}x + \frac{5}{4} = 0$

Score: ....

Answer Key

1) $y = x + 7$	2) $y = -\frac{5}{3}x - \frac{14}{3}$
3) $y = \frac{3}{2}x + \frac{1}{2}$	4) $y = \frac{7}{4}x + \frac{25}{4}$
5) $y = 6x - 9$	6) $y = -4x + 5$
7) $y = 6x + 9$	8) $y = -7x - 12$
9) $y = \frac{2}{3}x - 5$	10) $y = -\frac{1}{3}x - \frac{5}{3}$

Point-slope form

- Using the slope m and a point (x_1, y_1) on the line, the equation of the line is:
$$(y - y_1) = m(x - x_1)$$

EXAMPLE:

Write the point-slope form of an equation of a line with a slope of 3 that passes through the point $(3, -2)$.

The slope is 3, so m = 3. We also know one point, so we know $x_1 = 3$ and $y_1 = -2$. Now we can substitute these values into the general point-slope equation.

$y - y_1 = m(x - x_1) \to y - (-2) = \frac{3}{8}(x - 3) \to y + 2 = \frac{3}{8}(x - 3)$

PRACTICES:

Write an equation in point–slope form for the line that passes through the given point with the slope provided.

1) $(2, -3), m = 4$	2) $(-7, 4), m = \frac{1}{5}$
3) $(0, -6), m = -2$	4) $(-a, b), m = m$
5) $(-9, 1), m = 3$	6) $(3, 0), m = -5$
7) $(-4, 11), m = \frac{1}{3}$	8) $(0, 11), (-2, 11)$
9) $\left(-\frac{1}{3}, 3\right), m = \frac{1}{5}$	10) $(0, 0), (1, -3)$

Score: ....

Answer Key

1) $y + 3 = 4(x - 2)$	2) $y - 4 = \frac{1}{5}(x + 7)$
3) $y + 6 = -2x$	4) $y - b = m(x + a)$
5) $y - 1 = 3(x + 9)$	6) $y = -5(x - 3)$
7) $y - 11 = \frac{1}{3}(x + 4)$	8) $y - 11 = 0$
9) $y - 3 = \frac{1}{5}\left(x + \frac{1}{3}\right)$	10) $y = -3x$

Name: ....

Equation of Parallel or Perpendicular lines

- Parallel lines: The two lines will never intersect, and their slopes are identical. The only difference between the two lines is the y-intercept.
$\begin{cases} y = m_1x + b_1 \\ y = m_2x + b_2 \end{cases}$: Then, $m_1 = m_2$ and $b_1 \ne b_2$.

- Perpendicular Lines do intersect. Their intersection forms a right, or 90° angle. The slope of one line is the negative reciprocal of the slope of the other line.
$\begin{cases} y = m_1x + b_1 \\ y = m_2x + b_2 \end{cases}$: then, $m_1 = -\frac{1}{m_2}$, or $m_1 \times m_2 = -1$

EXAMPLE:

Given the functions below, identify the functions whose graphs are a pair of parallel lines and a pair of perpendicular lines. $\begin{cases} f(x) = 4x - 5 \\ h(x) = \frac{1}{4}x - 5 \end{cases}$ $\quad g(x) = -4x + 2$ $\quad p(x) = 4x + 7$

Parallel lines have the same slope ($f(x)$, and $p(x)$)

Perpendicular lines have negative reciprocal slopes ($g(x)$, and $h(x)$).

PRACTICES:

Write an equation of the line that passes through the given point and is parallel to the given line.	Write an equation of the line that passes through the given point and is perpendicular to the given line.
1) $(0,7), -5x - y = -4$	6) $(\frac{3}{5}, \frac{2}{5}), y = -6x - 24$
2) $(-2,-1), y = \frac{4}{5}x + 3$	7) $(-10, 0), y = \frac{5}{3}x - 15$
3) $(-2,5), -8x + 5y = -18$	8) $(3, -5), y = x + 12$
4) $(3,-2), y = -\frac{2}{5}x - 3$	9) $(-3,-1), y = \frac{7}{3}x - 4$
5) $(-5,-5), 6x + 15y = -30$	10) $(0,0), y - 8x + 6 = 0$

Score: ....

Answer Key

1) $y = -5x + 7$	2) $y = \frac{4}{5}x + \frac{3}{5}$
3) $y = \frac{8}{5}x + \frac{41}{5}$	4) $y = -\frac{2}{5}x - \frac{4}{5}$
5) $y = -\frac{2}{5}x - 7$	6) $y = \frac{1}{6}x + \frac{3}{10}$
7) $y = -\frac{3}{5}x - 6$	8) $y = -x - 2$
9) $y = -\frac{3}{7}x - \frac{16}{7}$	10) $y = -\frac{1}{8}x$

Equation of Horizontal and Vertical Lines

- ✓ Equations of horizontal and vertical lines only have one variable.
- ✓ The slope of horizontal lines is 0 and y-values for each point are the same. Then, the equation of horizontal lines is: $y = b$.
- ✓ The slope of vertical lines is undefined and the equation for a vertical line is: $x = a$

EXAMPLE:

Write an equation for the vertical line that passes through $(4, -1)$.

As the line is vertical, x is constant, and x always takes the same value. Then x always takes the value 4. Thus, the equation is $x = 4$.

PRACTICES:

Sketch the graph of each line.

1) $y = 3$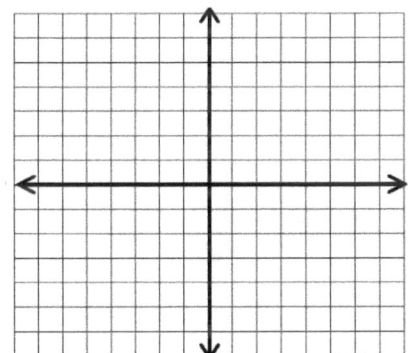

2) $y = -1$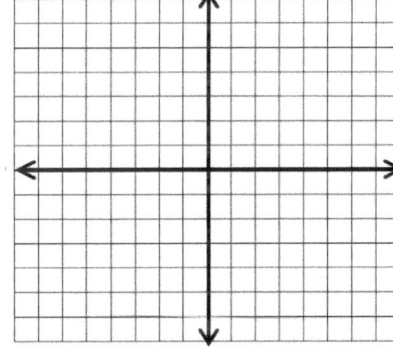

3) $x = 0$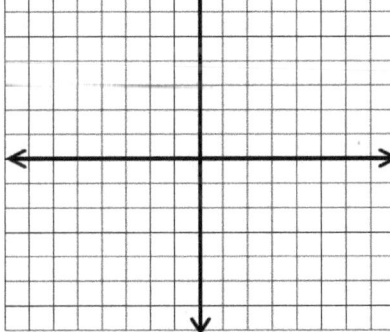

4) $x = 3$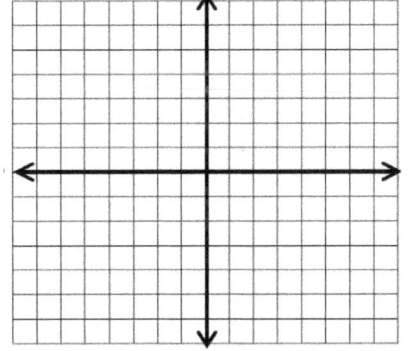

Answer Key

1) $y = 3$

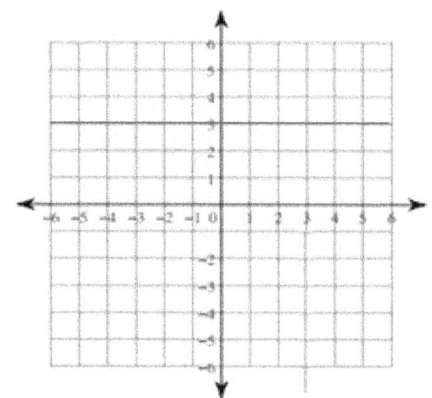

2) $y = -1$

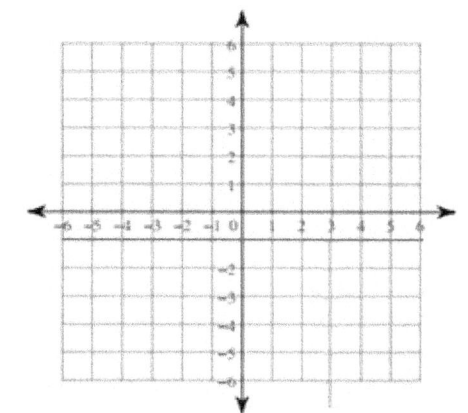

3) $x = 0$

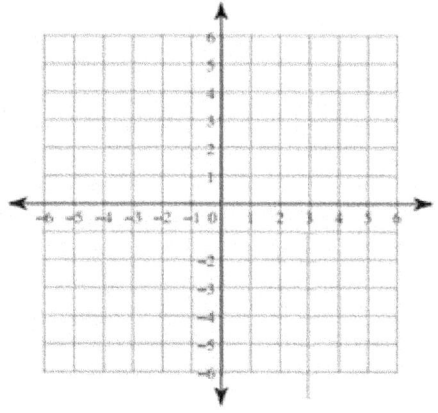

4) $x = 3$

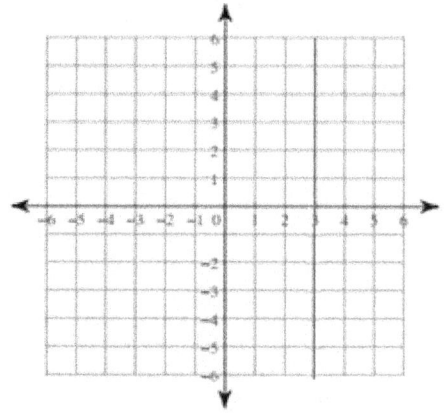

TASC Math Prep

Name: ....

Function Notation

- ✓ **Function notation** is a way to write functions that is easy to read and understand. Also, it determines that a relationship is a function.
- ✓ The notation y = f (x) defines a function named f. This is read y is a function of x. The letter x is the input value or independent variable. The letter y, or f (x), is the output value or dependent variable.

EXAMPLE:

Write $y = x^2 + 3x - 5$ using function notation and evaluate the function at $x = 2$.

By applying function notation, we get: $f(x) = x^2 + 3x - 5$

Evaluation: Substitute x with 2:

$f(2) = 2^2 + 3 \times 2 - 5 = 4 + 6 - 5 = 5$

PRACTICES:

Write in function notation.

1) $v = 8t$	2) $r = 4p^2 + 2p - 2$
3) $h = 15g + 8$	4) $y = 5x - \frac{3}{4}$

Evaluate each function.

5) $h(x) = x^3 - 8$, find $h(-2)$	6) $f(u) = 9u - 2$, find $f(1/3)$
7) $h(x) = 3x - 6$, find $h(a)$	8) $h(a) = -2a + 4$, find $h(3b)$
9) $h(x) = x^2 + 4x - 7$, find $h(x^2)$	10) $h(x) = x^2 + 5$, find $h(-\frac{a}{3})$

www.mathnotion.com

Score: ....

Answer Key

1) $v(t) = 8t$	2) $r(p) = 4p^2 + 2p - 2$
3) $h(g) = 15g + 8$	4) $f(x) = 5x - \frac{3}{4}$
5) -16	6) 1
7) $3a - 6$	8) $-6b + 4$
9) $x^4 + 4x^2 - 7$	10) $\frac{1}{9}a^2 + 5$

Adding and Subtracting Functions

✓ Like numbers and polynomials, we can add and subtract functions which results into a new function.

✓ Let f(x) and g(x) be two functions:
We can add two functions as: $(f + g)(x) = f(x) + g(x)$
We can subtract two functions as: $(f - g)(x) = f(x) - g(x)$

EXAMPLE:

Find the sum, and difference of $f = 3x + 1$ and $g = x - 2$ at the point 5.

$f + g = (3x + 1) + (x - 2) = 4x - 1 \rightarrow (f + g)(5) = 19$
$f - g = (3x + 1) - (x - 2) = 2x + 3 \rightarrow (f - g)(5) = 13$

PRACTICES:

Perform the indicated operation.

1) $h(t) = 5t - 3$
   $g(t) = 5t + 3$
   Find $(h - g)(t)$.

2) $h(n) = 4n - 4$
   $g(n) = n^2 - 6n + 9$
   Find $(h + g)(a)$.

3) $g(a) = -3a^2 + 4$
   $f(a) = 2a^2 - a + 4$
   Find $(g - f)(a)$.

4) $g(x) = -x^2 + 8 - 3x$
   $f(x) = 8 + 2x$
   Find $(g - f)(x)$.

5) $h(x) = 3x^2 - 5$
   $g(x) = -4x^2 + 2x$
   Find $(h + g)(t)$.

6) $g(t) = t + 8$
   $f(t) = -3t^2 + t$
   Find $(g - f)(u - 1)$.

7) $h(x) = -3x + 4$
   $g(x) = 2x - 6$
   Find $(h + g)(2)$.

8) $k(x) = -3x + 6$
   $h(x) = x^2 + 2x + 4$
   Find $(k + h)(t - 3)$.

Score: ....

Answer Key

1) -6	2) $a^2 - 2a + 5$
3) $-5a^2 + a$	4) $-x^2 - 5x$
5) $-t^2 + 2t - 5$	6) $3u^2 - 6u + 11$
7) -4	8) $t^2 - 7t + 22$

Name: ....

Multiplying and Dividing Functions

- ✓ Like numbers and polynomials, we can multiply and divide functions which results into a new function.
- ✓ Let f(x) and g(x) be two functions:
  We can multiply two functions as: $(f \cdot g)(x) = f(x) \cdot g(x)$
  We can divide two functions as: $\left(\frac{f}{g}\right)(x) = \frac{f(x)}{g(x)}$

EXAMPLE:

Given that $f(x) = x + 3$ and $g(x) = x^2 - 9$, find $(fg)(x)$ and $\left(\frac{f}{g}\right)(x)$ at the point 1.

$(fg)(x) = f(x) \times g(x) = (x+3)(x^2-9) = x^3 + 3x^2 - 9x - 27 \rightarrow (fg)(1) = -32$

$\left(\frac{f}{g}\right)(x) = \frac{f(x)}{g(x)} = \frac{x+3}{x^2-9} = \frac{x+3}{(x-3)(x+3)} = \frac{1}{x-3} \rightarrow \left(\frac{f}{g}\right)(1) = -\frac{1}{2}$

PRACTICES:

Perform the indicated operation.

1) $f(x) = x^2 - 3x$ $g(x) = 4x^2 - 2$ Find $(f.g)(x)$	2) $g(t) = \frac{1}{3}t^2 + \frac{1}{3}$ $h(t) = 3t - 3$ Find $(h.g)(\frac{1}{3})$
3) $f(a) = 12a - 8$ $g(a) = 5a + 10$ Find $\left(\frac{f}{g}\right)(-1)$	4) $h(a) = -2a$ $g(a) = -6a^2 - 2a$ Find $\left(\frac{h}{g}\right)(a)$
5) $g(a) = 4a - 3$ $h(a) = 4a - 2$ Find $(g.h)(2)$	6) $k(n) = 2n^2 - n$ $h(n) = 3n^2 - 3$ Find $(k.h)(1)$
7) $f(x) = 2x + 1$ $g(x) = 4x^2 - 1$ Find $\left(\frac{g}{f}\right)(x)$	8) $f(t) = -a + 3$ $g(t) = a^3 + 2$ Find $\left(\frac{3f}{g}\right)(a)$

Score: ....

Answer Key

1) $4x^4 - 12x^3 - 2x^2 + 6x$	2) $-\dfrac{20}{27}$
3) -4	4) $\dfrac{1}{3a+1}$
5) 30	6) 0
7) $2x - 1$	8) $\dfrac{-3a+9}{a^3+2}$

Composition of Functions

- A composite function is generally a function that is written inside another function. Composition of a function is done by substituting one function into another function.
- The notation used for composition is:

$$(f \circ g)(x) = f(g(x))$$

EXAMPLE:

Using $f(x) = x + 1$ and $g(x) = 2x$, find: $(f \circ g)(1)$

$(f \circ g)(x) = f(g(x)) = 2(x + 1) = 2x + 2$
$(f \circ g)(1) = 4$

PRACTICES:

Using $f(x) = 2x - 8$, and $g(x) = -2x + 1$, find:

1) $f(g(0))$
2) $f(f(1))$
3) $g(f(3))$

Using $f(x) = 3x - 4a$, and $g(x) = x^2 - 2$, find:

4) $(f \circ g)(1) = f(g(1))$
5) $(f \circ f)(3)$
6) $(g \circ f)(2)$

Using $f(x) = -2x + 3$, and $g(x) = x - b$, find:

7) $(f \circ g)(-2x)$
8) $(f \circ g)(x + 1)$
9) $(g \circ f)(x^2)$

Score: ....

Answer Key

1) -6	2) -20
3) 5	4) $-3 - 4a$
5) $27 - 16a$	6) $16a^2 - 48a + 34$
7) $4x + 3 + 2b$	8) $-2x + 1 + 2b$
9) $-2x^2 + 3 - b$	

Name: ....

Solve a Quadratic Equation

- ✓ Write the equation in the Standard form:
  $ax^2 + bx + c = 0$ (One side must only contain zero)
- ✓ Factorize the quadratic.
- ✓ Use quadratic formula if you couldn't factorize the quadratic.
- ✓ Quadratic formula: $x = \frac{-b \pm \sqrt{b^2 - 4ac}}{2a}$

EXAMPLE:

Solve: $x^2 - 4x - 21 = 0$.

$\begin{cases} a = 1 \\ b = -4 \\ c = -21 \end{cases} \Rightarrow x = \frac{-b \pm \sqrt{b^2 - 4ac}}{2a} = \frac{-(-4) \pm \sqrt{(-4)^2 - 4(1)(-21)}}{2(1)} = \begin{cases} \frac{4 + \sqrt{100}}{2} = 7 \\ \frac{4 - \sqrt{100}}{2} = -3 \end{cases}$

This equation is also factorable:

$x^2 - 4x - 21 = 0 \rightarrow (x - 7)(x + 3) = 0 \rightarrow \begin{cases} x - 7 = 0 \rightarrow x = 7 \\ x + 3 = 0 \rightarrow x = -3 \end{cases}$

PRACTICES:

Solve each equation by using the quadratic formula.

1) $x^2 + 6x = -8$	2) $3x^2 - 9x - 9 = 3$
3) $6x^2 = 24x - 18$	4) $x^2 = 3x$
5) $3x^2 + 45 = -24x$	6) $2x^2 - 24x = -72$
7) $-6x^2 - 10x - 4 = 8 - 4x^2$	8) $x^2 - 20x = -84$
9) $2x^2 + 18x + 52 = 12$	10) $x^2 + 2x = 15 + 4x$

Score: ....

Answer Key

1) {–4, –2}	2) {4, –1}
3) {1, 3}	4) {3, 0}
5) {– 3, –5}	6) {6}
7) {–3, –2}	8) {14, 6}
9) {–4, –5}	10) {5, – 3}

Chapter 9 : Geometry

Topics that you'll learn in this chapter:

- The Pythagorean Theorem
- Area of Triangles and Trapezoids
- Area and Circumference of Circles
- Area and Perimeter of Polygons
- Area of Squares, Rectangles, and Parallelograms
- Volume of Cubes, Rectangle Prisms, and Cylinder
- Surface Area of Cubes, Rectangle Prisms, and Cylinder

"Mathematics is, as it were, a sensuous logic, and relates to philosophy as do the arts, music, and plastic art to poetry." — *K. Shegel*

The Pythagorean Theorem

- In any right triangle: $a^2 + b^2 = c^2$

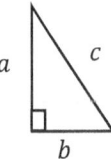

EXAMPLE:

Find the missing length.

Use Pythagorean Theorem: $a^2 + b^2 = c^2$

Then: $a^2 + b^2 = c^2 \to 3^2 + 4^2 = c^2 \to 9 + 16 = c^2$

$c^2 = 25 \to c = 5$

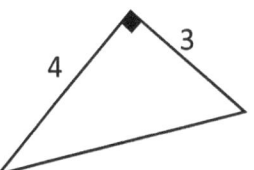

PRACTICES:

Do the following lengths form a right triangle?	Find each missing length to the nearest tenth.
1) sides 3, 4, 5	2) legs 32 and 60, hypotenuse ?
3) sides 13, 11, $\sqrt{290}$	4) leg ?, leg 35, hypotenuse 72
5) sides 9, 12, 25	6) legs 55 and 20, hypotenuse ?
7) sides 5, 12, 13	8) leg 48, leg 64, hypotenuse ?

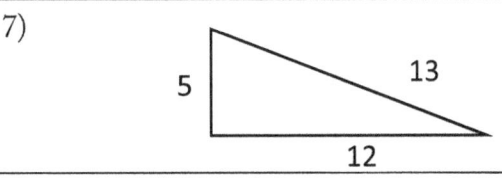
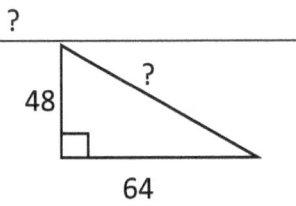

9) Triangle with sides 6, 14, 22

10) Right triangle with hypotenuse 13, one leg 5, and other leg ?

Score: ....

Answer Key

1) Yes	2) 68
3) Yes	4) 62.92
5) No	6) 58.52
7) Yes	8) 80
9) No	10) 12

Angles

- ✓ **Adjacent:** Two triangles are said to be adjacent if its two angles have common side, common vertex and do not overlap.

- ✓ **Vertical:** Two angles share same vertex.

- ✓ **Complementary:** Sum of the measure of two complementary angles are 90°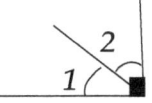

- ✓ **Supplementary:** Sum of the measure of two supplementary angles are 180°

EXAMPLE:

Find x.

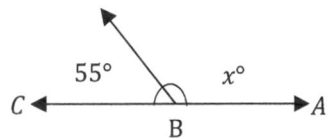

Supplementary: Sum of the measure of two supplementary angles are 180°.

Then: $180° - 55° = 125°$

PRACTICES:

What is the value of x in the following figures?

1)

2)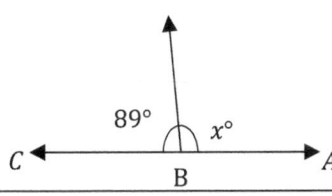

3)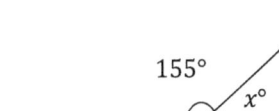

4)

5)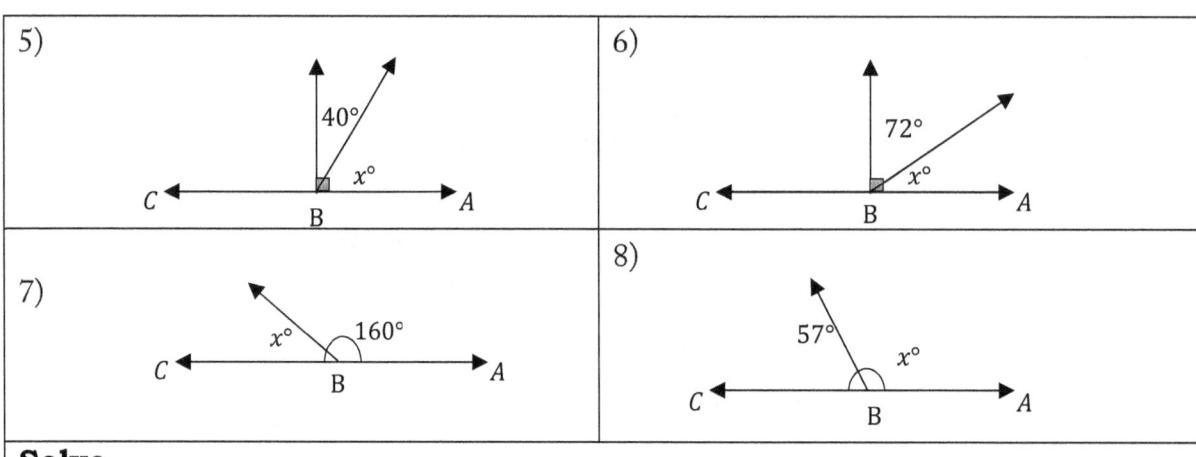

6)

7)

8)

Solve.

9) Six supplement peer to each other angles have equal measures. What is the measure of each angle? _____

10) The measure of an angle is one fourth the measure of its complementary. What is the measure of the angle? _____

Score: ....

Answer Key	
1) 60°	2) 91°
3) 32°	4) 25°
5) 50°	6) 18°
7) 20°	8) 123°
9) 30°	10) 18°

Area of Triangles

- In any triangle the sum of all angles is 180 degrees.
- Area of a triangle = $\frac{1}{2}$ (base × height)

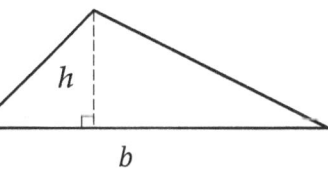

EXAMPLE:

What is the area of triangle?

Solution:

Use the are formula: Area = $\frac{1}{2}$ (base × height)

base = 12 and height = 8

Area = $\frac{1}{2}(12 \times 8) = \frac{1}{2}(96) = 48$

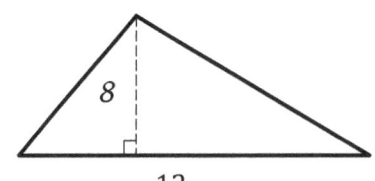

PRACTICES:

Find the area of each.

1) c = 15 mi
   h = 4 mi

2) c = 6 m
   h = 5.2 m

3) a = 9.5 m
   b = 25 m
   c = 18 m
   h = 9 m
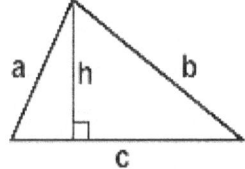

4) s = 8 m
   h = 6.93 m
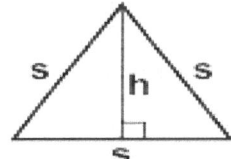

5) c = 25 mi
   h = 8 mi
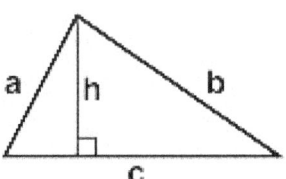

6) c = 10 m
   h = 6.4 m
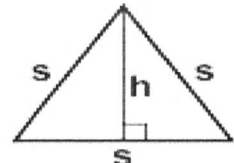

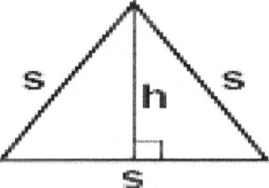

www.mathnotion.com

TASC Math Prep

7) a = 3.5 m b = 7 m c = 16 m h = 7 m 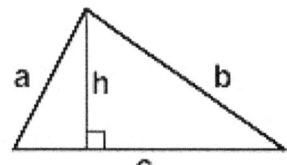	8) s = 12 m h = 4.64 m 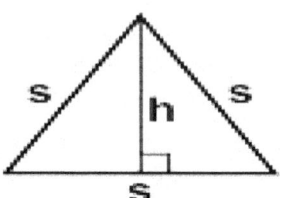
9) c = 13 mi h = 6 mi	10) S = 18 m h = 7.4 m 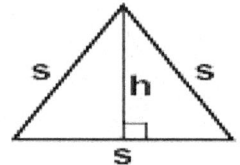

Score: ....

Answer Key

1) 30 mi²	2) 15.6 m²
3) 81 m²	4) 27.72 m²
5) 100 mi²	6) 32 m²
7) 56 m²	8) 27.84 m²
9) 39 mi²	10) 133.2 m²

Area of Trapezoids

- A trapezoid is a quadrilateral with at least one pair of parallel sides.
- Area of a trapezoid = $\frac{1}{2}h(b_1 + b_2)$

EXAMPLE:

Calculate the area of the trapezoid.

Use area formula: $A = \frac{1}{2}h(b_1 + b_2)$

$b_1 = 12$, $b_2 = 16$ and $h = 18$

Then: $A = \frac{1}{2}18(12 + 16) = 9(28) = 252\ cm^2$

PRACTICES:

Calculate the area for each trapezoid.

1)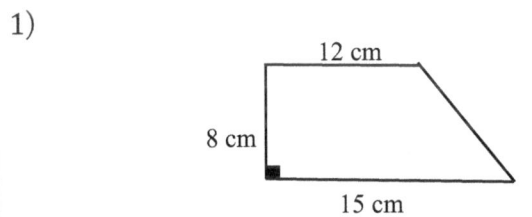

2)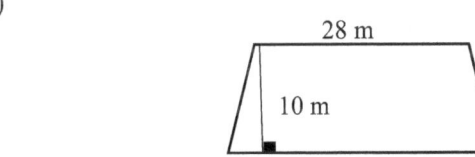

3)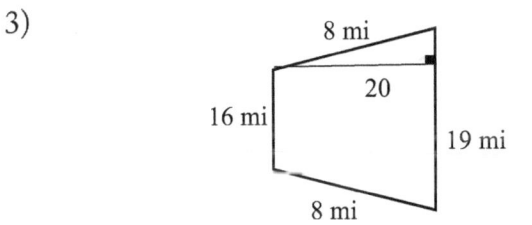

4)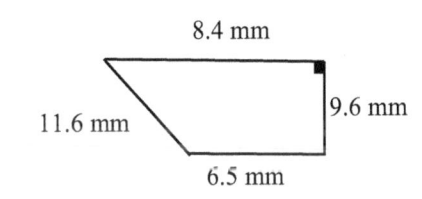

5)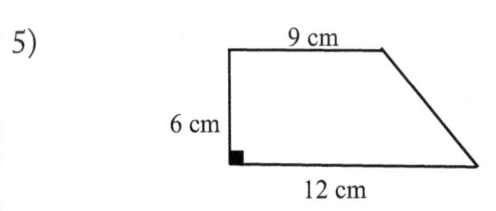

6)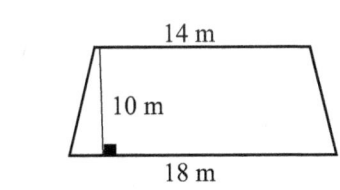

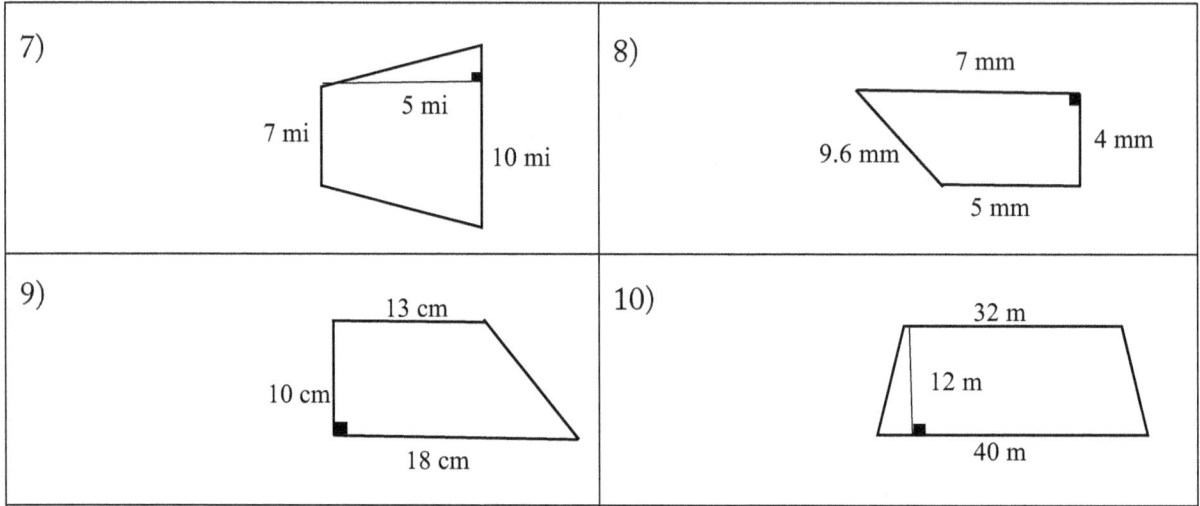

Score: ....

Answer Key

1) 108 cm²	2) 290 m²
3) 350 mi²	4) 71.52 mm²
5) 63 cm²	6) 160 m²
7) 42.5 mi²	8) 24 mm²
9) 155 cm²	10) 432 m²

TASC Math Prep

Name: ....

Area and Perimeter of Polygons

Perimeter of a square
= 4 × side = 4s

Perimeter of a rectangle
= 2(width + length)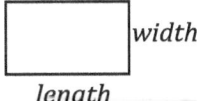

Perimeter of trapezoid
= a + b + c + d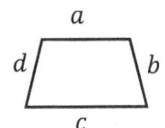

Perimeter of a regular hexagon = 6a

Perimeter of a parallelogram = 2(l + w)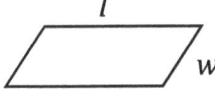

EXAMPLE:

Find the perimeter of following regular hexagon.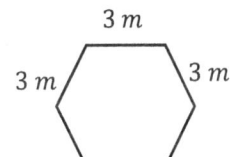

Perimeter of Pentagon = 6a

Perimeter of Pentagon = 6a = 6 × 3 = 18m

PRACTICES:

Find the area and perimeter of each.

1)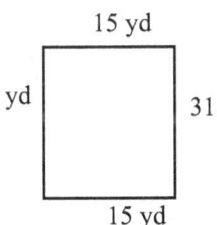

2)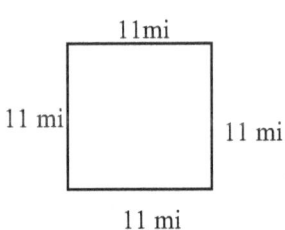

3)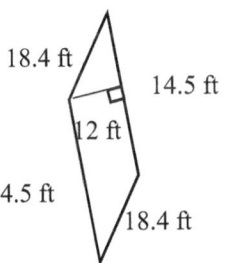

4)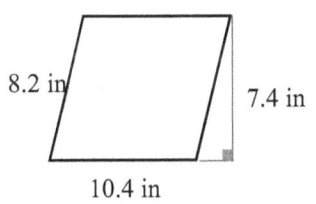

www.mathnotion.com 215

TASC Math Prep

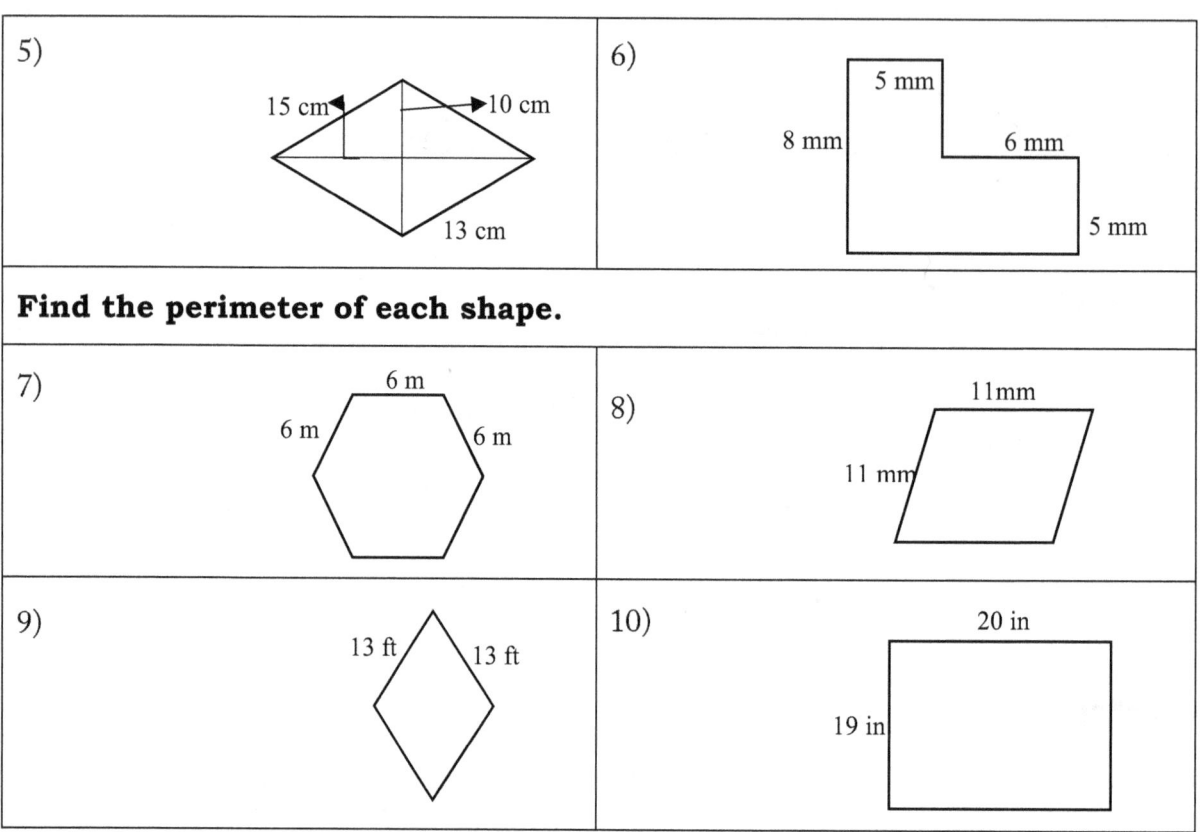

Find the perimeter of each shape.

Score: ....

Answer Key

1) Area: 472.5 yd^2, Perimeter: 93 yd	2) Area: 121 mi^2, Perimeter: 44 mi
3) Area: 174 ft^2, Perimeter: 65.8 ft	4) Area: 76.96 in^2, Perimeter: 37.2 in
5) Area: 75 cm^2, Perimeter 52 cm	6) Area: 70 mm^2, Perimeter: 38 mm
7) P: 36 m	8) P: 44 mm
9) P: 52 ft	10) P: 78 in

www.mathnotion.com

Area and Circumference of Circles

- We use variable r for the radius and d for diameter in a circle and π is about 3.14.
- Area of a circle $= \pi r^2$
- Circumference of a circle $= 2\pi r$

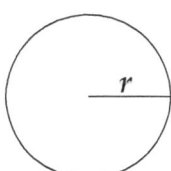

EXAMPLE:

Find the Circumference and area of the circle.

Use Circumference formula: $Circumference = 2\pi r$

$r = 6$, then: $Circumference = 2\pi(6) = 12\pi$

$\pi = 3.14$ then: $Circumference = 12 \times 3.14 = 37.68$

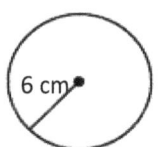

Use area formula: $Area = \pi r^2$,

$r = 6$ then: $Area = \pi(6)^2 = 36\pi$, $\pi = 3.14$ then: $Area = 36 \times 3.14 = 113.04$

PRACTICES:

Find the area and circumference of each. ($\pi = 3.14$)

1)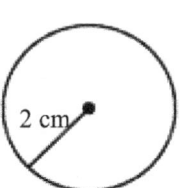

2)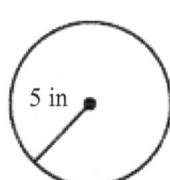

3)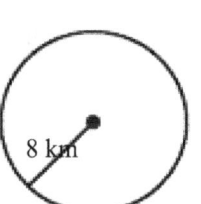

4)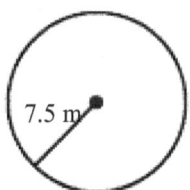

TASC Math Prep

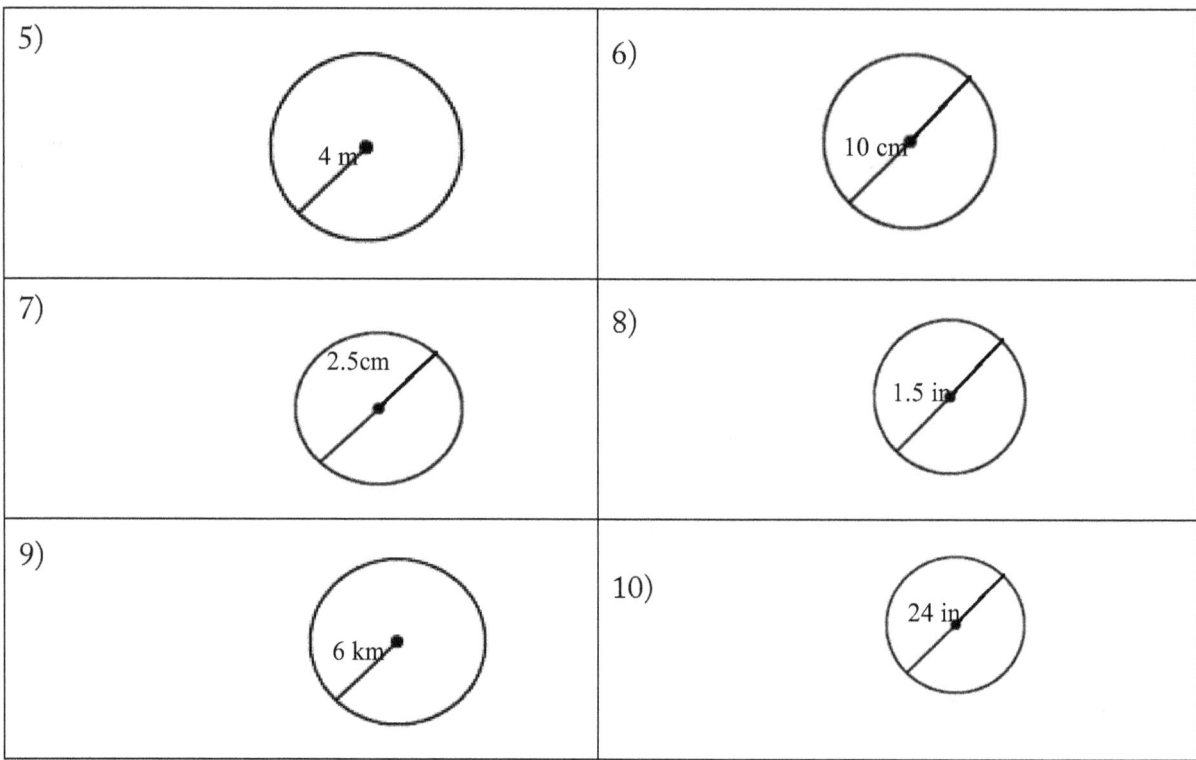

Score: ....

Answer Key

1) Area: 12.56 cm², Circumference: 12.56cm.	2) Area: 78.5 in2, Circumference: 31.4 in.
3) Area: 200.96 km², Circumference: 50.24 km.	4) Area: 176.625 m2, Circumference: 47.1 m.
5) Area: 50.24 m², Circumference: 25.12 m	6) Area: 78.5 cm2, Circumference: 31.4 cm.
7) Area: 4.906 cm², Circumference: 7.85 cm.	8) Area: 1.766 in2, Circumference: 4.71 in.
9) Area: 113.04 km², Circumference:37.68 km.	10) Area: 452.16 in², Circumference: 75.36 in

Volume of Cubes

- ✓ A three-dimensional solid object bounded by six square sides is called cube.
- ✓ The measure of the amount of space inside of a solid figure is called volume, like a cube, ball, cylinder, or pyramid.
- ✓ Volume of a cube = $(one\ side)^3$

EXAMPLE:

Find the volume of this cube.

Use volume formula: $volume = (one\ side)^3$

Then: $volume = (one\ side)^3 = (2)^3 = 8\ cm^3$

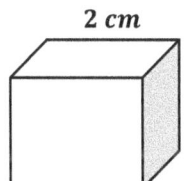
2 cm

PRACTICES:

Find the volume of each.

1)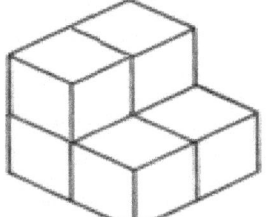

2)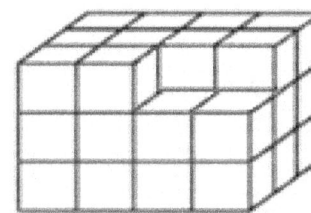

3)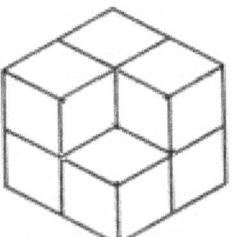

4)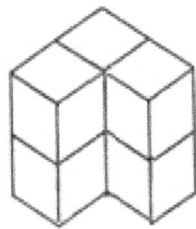

5)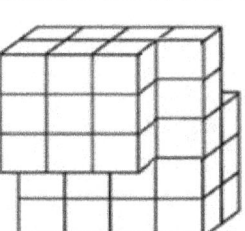

6)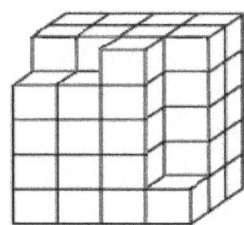

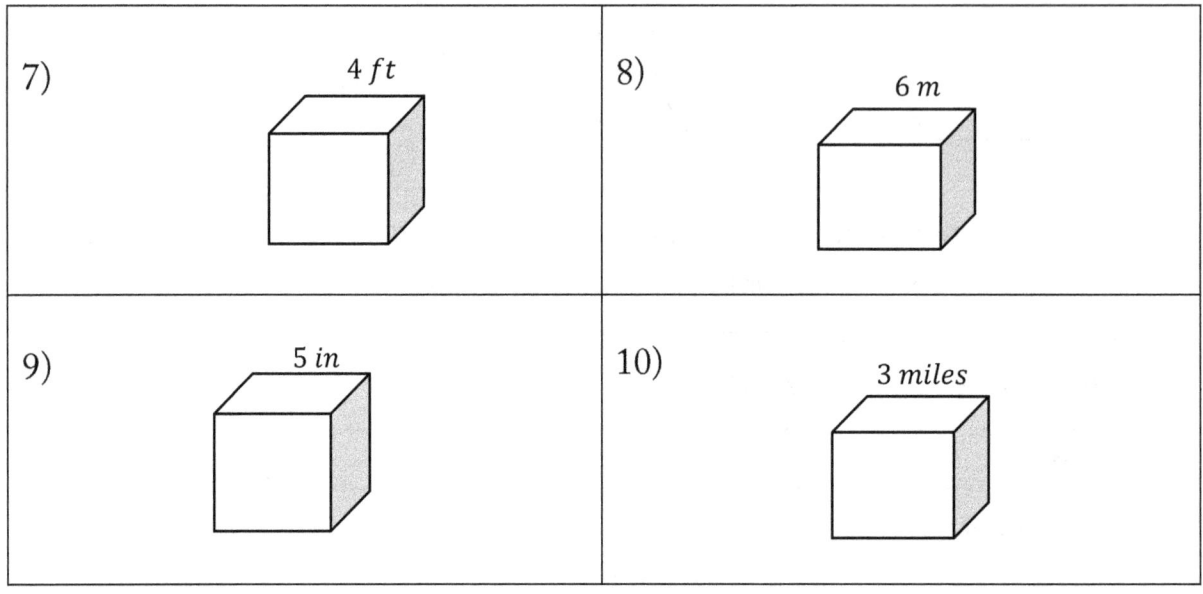

7) 4 ft
8) 6 m
9) 5 in
10) 3 miles

Score: ....

Answer Key

1) 6	2) 34
3) 7	4) 6
5) 41	6) 54
7) $64\ ft^3$	8) $216\ m^3$
9) $125\ in^3$	10) $27\ mi^3$

Volume of Rectangle Prisms

- ✓ A solid 3-dimensional object which has six rectangular faces.
- ✓ Volume of a Rectangular prism = **Length × Width × Height**

  Volume = $l \times w \times h$

EXAMPLE:

Find the volume and surface area of rectangular prism.

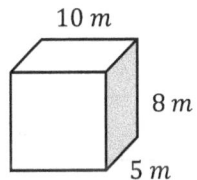

Use volume formula: $Volume = l \times w \times h$

Then: $Volume = 10 \times 5 \times 8 = 400 \, m^3$

PRACTICES:

Find the volume of each of the rectangular prisms.

1)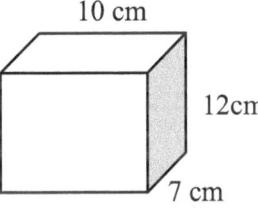

2)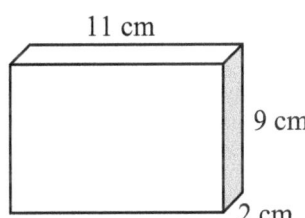

3)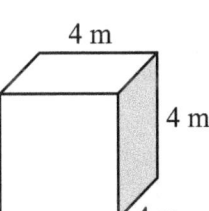

4)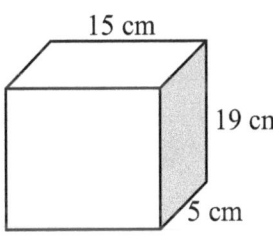

5)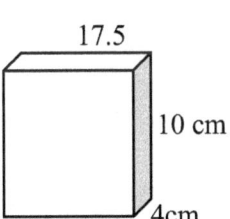

6)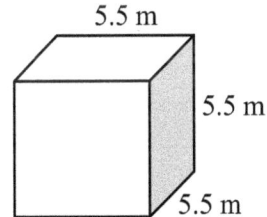

www.mathnotion.com

TASC Math Prep

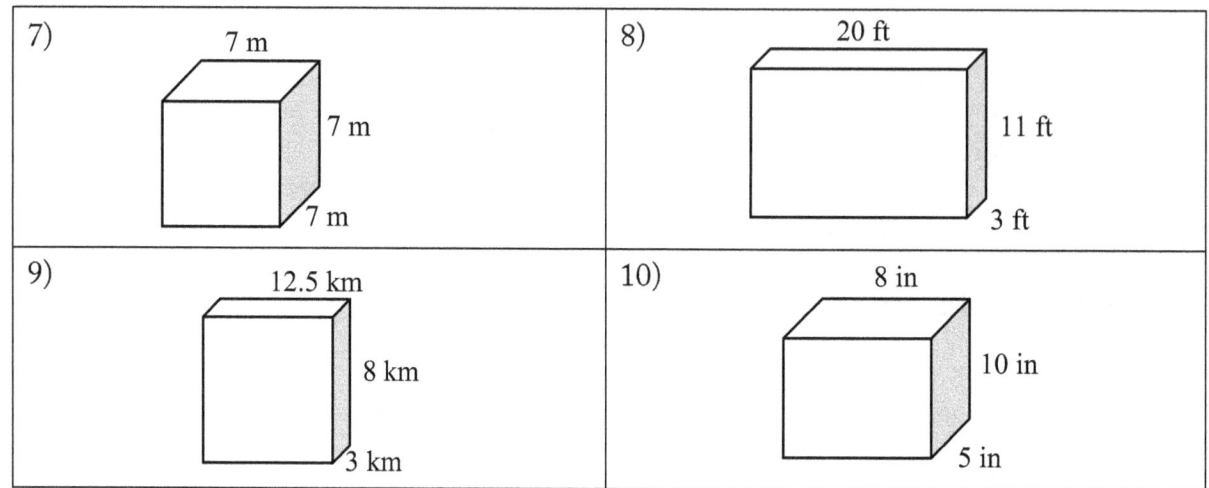

Score: ....

Answer Key

1) 840 cm³	2) 198 cm³
3) 64 m³	4) 1,425 cm³
5) 700 cm³	6) 166.375 cm³
7) 343 cm³	8) 660 ft³
9) 300 km³	10) 400 in³

Surface Area of Cubes

- A three-dimensional solid object bounded by six square sides is called cube.

  surface area of cube = 6 × (one side)²

EXAMPLE:

Find the volume and surface area of this cube.

surface area of cube: $6(one\ side)^2 = 6(2)^2 = 6(4) = 24\ cm^2$

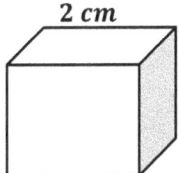

PRACTICES:

Find the surface of each cube.

1) 7 mm

2) 10.5 mm

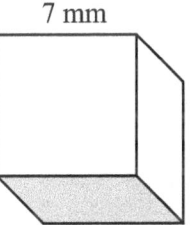 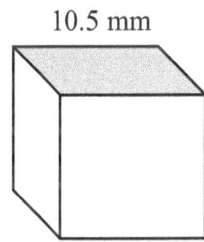

3) 3.5 cm

4) 4 m

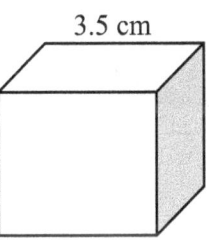 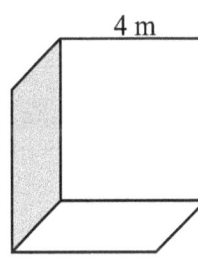

5)

6) 8.1 ft

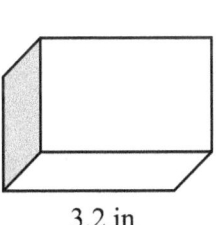

3.2 in

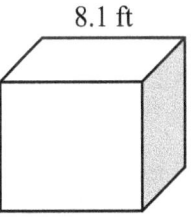

TASC Math Prep

7) 1.6 in	8) 11 m
9) 5.2 in	10) 2.25 mm

Score: ....

Answer Key

1) 294 mm²	2) 661.5 mm²
3) 73.5 cm²	4) 96 m²
5) 61.44 in²	6) 393.66 ft²
7) 15.36 in²	8) 726 m²
9) 162.24 in²	10) 30.375 mm²

Surface Area of a Rectangle Prism

- A solid 3-dimensional object which has six rectangular faces.
  Surface area= $2(wh + lw + lh)$

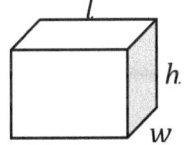

EXAMPLE:

Find the volume and surface area of rectangular prism.

Use surface area formula: $Surface\ area = 2(wh + lw + lh)$

Then: $Surface\ area = 2(5 \times 8 + 10 \times 5 + 10 \times 8) =$

$2(40 + 50 + 80) = 340\ m^2$

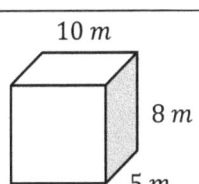

PRACTICES:

Find the surface of each prism.

1)

3 yd, 4 yd, 6 yd

2)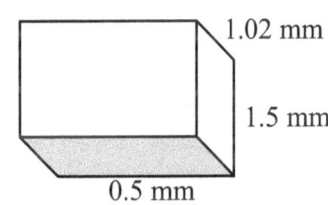

3)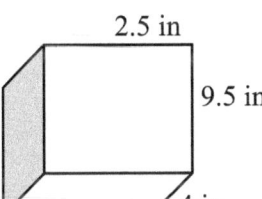

4)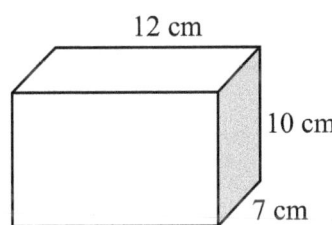

5)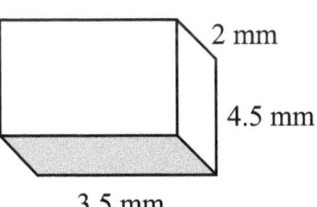

6)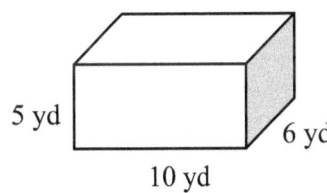

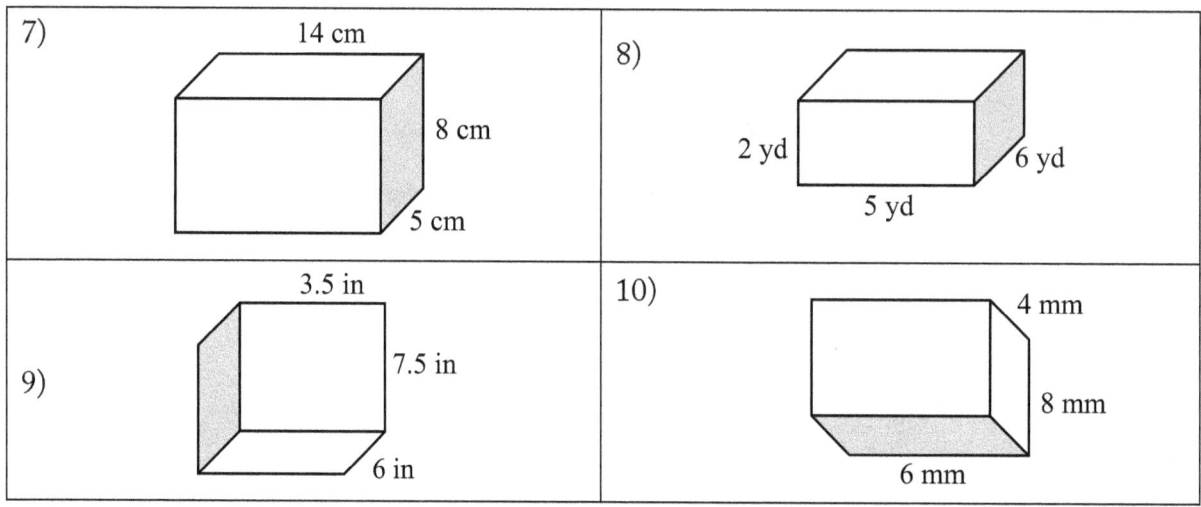

Score: ....

Answer Key

1) 108 yd²	2) 5.58 mm²
3) 143.5 in²	4) 548 cm²
5) 63.5 mm²	6) 280 yd²
7) 444 cm²	8) 104 yd²
9) 184.5 in²	10) 208 mm²

TASC Math Prep

Name: ....

Volume of a Cylinder

- ✓ A solid geometric figure with straight parallel sides and a circular cross section is called a cylinder.
- ✓ Volume of Cylinder Formula = π(radius)² × height
  π = 3.14

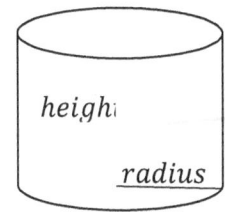

EXAMPLE:

Find the volume of the follow Cylinder.

Use volume formula: $Volume = \pi(radius)^2 \times height$

Then: $Volume = \pi(4)^2 \times 6 = \pi 16 \times 6 = 96\pi$

π = 3.14 then: Volume = 96π = 301.44

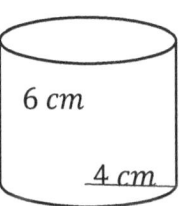

PRACTICES:

Find the volume of each cylinder. (π = 3.14).

1)

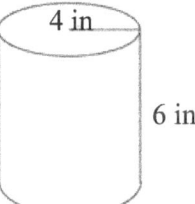

2)

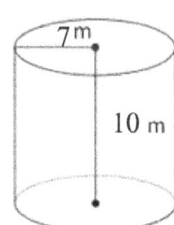

3)

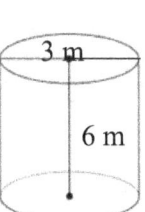

4)

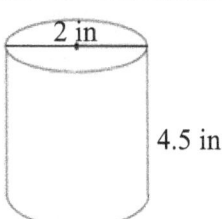

5)

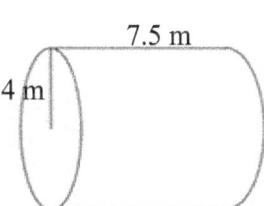

6)

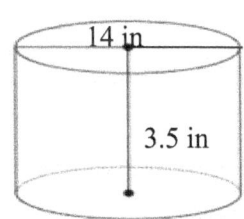

www.mathnotion.com

TASC Math Prep

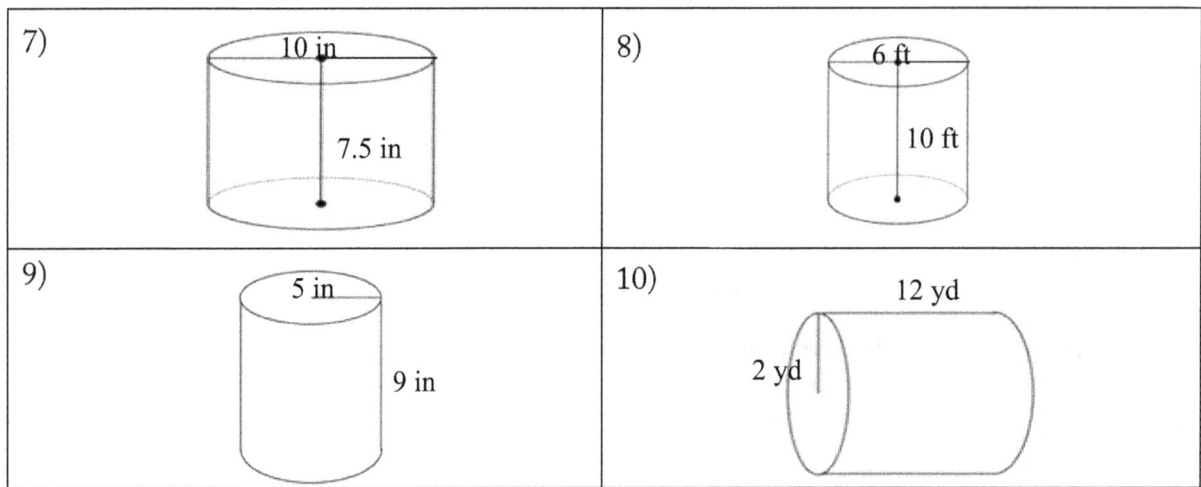

7) [cylinder: 10 in diameter, 7.5 in height]
8) [cylinder: 6 ft diameter, 10 ft height]
9) [cylinder: 5 in diameter, 9 in height]
10) [cylinder: 12 yd length, 2 yd radius]

Score: ....

Answer Key	
1) 301.44 in³	2) 1538.6 m³
3) 42.39 m³	4) 14.13 in³
5) 376.8 m³	6) 538.51 in³
7) 588.75 in³	8) 282.6 ft³
9) 706.5 in³	10) 150.72 yd³

TASC Math Prep

Name: ....

Surface Area of a Cylinder

- ✓ A solid geometric figure with straight parallel sides and a circular cross section is called cylinder.
- ✓ Surface area of a cylinder = $2\pi r^2 + 2\pi rh$

EXAMPLE:

Find the Surface area of the follow Cylinder.

Use surface area formula: $Surface\ area = 2\pi r^2 + 2\pi rh$

Then: $= 2\pi(4)^2 + 2\pi(4)(6) = 2\pi(16) + 2\pi(24) = 32\pi + 48\pi = 80\pi$

$\pi = 3.14$ then: $Surface\ area = 80 \times 3.14 = 251.2$

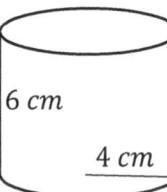

PRACTICES:

Find the surface of each cylinder. ($\pi = 3.14$).

1)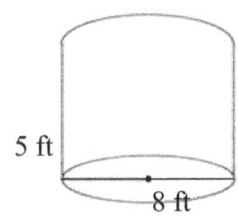
   5 ft, 8 ft

2)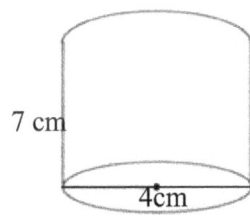
   7 cm, 4 cm

3)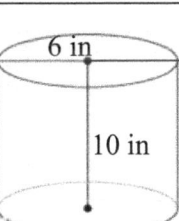
   6 in, 10 in

4)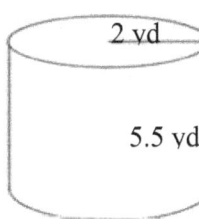
   2 yd, 5.5 yd

5)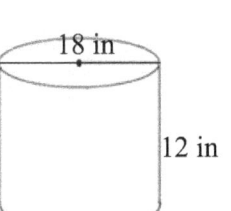
   18 in, 12 in

6)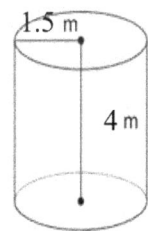
   1.5 m, 4 m

www.mathnotion.com

TASC Math Prep

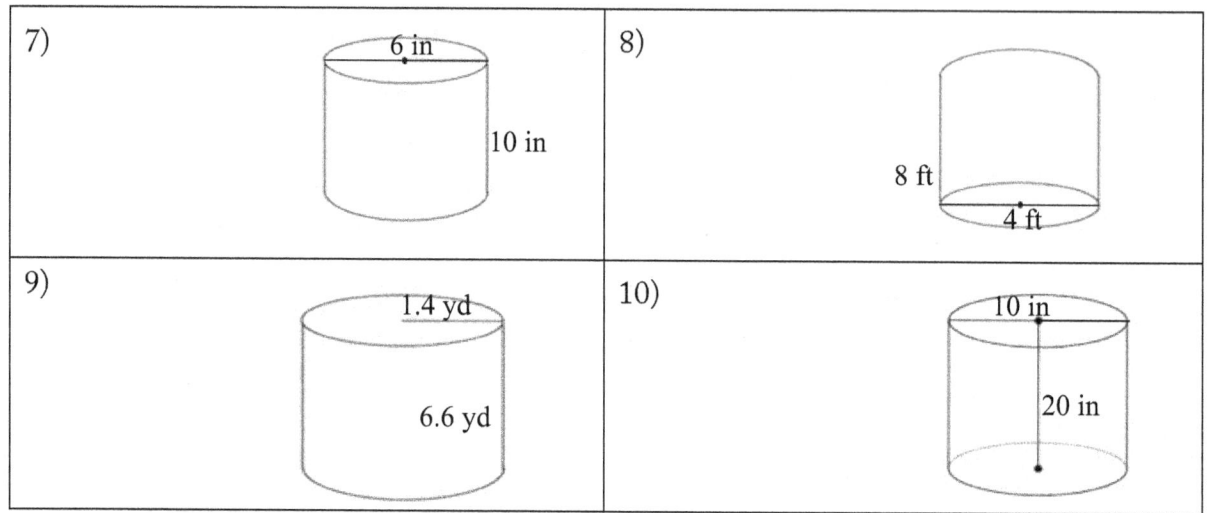

Score: ....

Answer Key

1) 226.08 ft²	2) 113.04 cm²
3) 224.92 in²	4) 94.2 yd²
5) 1,186.92 in²	6) 51.81 m²
7) 244.92 in²	8) 502.4 ft²
9) 70.336 yd²	10) 785 in²

www.mathnotion.com

Chapter 10 : Statistics

Topics that you'll learn in this chapter:

- Mean, Median, Mode, and Range of the Given Data
- Box and Whisker Plots
- Bar Graph
- Stem- And- Leaf Plot
- The Pie Graph or Circle Graph
- Dot and Scatter Plots
- Probability of Simple Events

"The book of nature is written in the language of Mathematic" -Galileo

Name: ....

Mean and Median

- **Mean:** $\dfrac{\text{sum of the data}}{\text{total number of data entires}}$
- **Median:** Middle value in the sorted list of numbers
- When there are two middle numbers, we average them

EXAMPLE:

What is the median of these numbers? 4, 9, 13, 8, 15, 18, 5, 11

Write the numbers in order: 4, 5, 8, 9, 11, 13, 15, 18

Median is the number in the middle. Therefore, there are 9 and 11 in the middle, then find the average: $\dfrac{9+11}{2} = \dfrac{20}{2} = 10$, the median is 10

PRACTICES:

Find Mean and Median of the Given Data.

1) 8, 10, 7, 3, 12	2) 4, 6, 9, 7, 5, 19
3) 5, 11, 1, 1, 8, 9, 20	4) 12, 4, 2, 7, 3, 2
5) 3, 5, 7, 4, 7, 8, 9	6) 5, 10, 4, 4, 9, 12, 9
7) 10, 4, 8, 5, 9, 6, 7, 19	8) 16, 3, 4, 3, 7, 6, 18

Solve.

9) In a javelin throw competition, five athletics score 23, 45, 53, 53, 13 and 61 meters. What are their Mean and Median? _____

10) Eva went to shop and bought 7 apples, 4 peaches, 6 bananas, 3 pineapples and 4 melons. What are the Mean and Median of her purchase? _____

Score: ....

Answer Key

1) Mean: 8, Median: 8	2) Mean: 8.33, Median: 6.5
3) Mean: 7.85, Median: 8	4) Mean: 5, Median: 3.5
5) Mean: 6.14, Median: 7	6) Mean: 7.57, Median: 9
7) Mean: 8.5, Median: 7.5	8) Mean: 8.14, Median: 6
9) Mean: 39.106, Median: 45	10) Mean: 4.8, Median: 4

Mode and Range

- ✓ Mode: The most appeared value in the list.
- ✓ Range: The difference of highest value and lowest value in the list

EXAMPLE:

What is the mode(s) of these numbers? **22, 16, 12, 9, 7, 6, 4, 6, 9**

Mode: The most appeared value in the list.

Therefore: modes are 6 and 9

PRACTICES:

Find Mode and Rage of the Given Data.

1) 10, 12, 8, 8, 4, 1, 9
   Mode: _____ Range: _____

2) 4, 6, 4, 13, 2, 13, 19, 13
   Mode: _____ Range: _____

3) 8, 8, 7, 2, 7, 7, 5, 6, 5
   Mode: _____ Range: _____

4) 12, 9, 12, 6, 12, 9, 10
   Mode: _____ Range: _____

5) 2, 2, 4, 3, 2, 10, 8
   Mode: _____ Range: _____

6) 6, 1, 4, 20, 19, 2, 7, 1, 5, 1
   Mode: _____ Range: _____

7) 16, 35, 9, 7, 7, 5, 14, 13, 7
   Mode: _____ Range: _____

8) 7, 6, 6, 9, 16, 6, 7, 5
   Mode: _____ Range: _____

Solve.

9) A stationery sold 15 pencils, 26 red pens, 22 blue pens, 10 notebooks, 12 erasers, 22 rulers and 42 color pencils. What are the Mode and Range for the stationery sells?
   Mode: _____ Range: _____

10) In an English test, eight students score 24, 13, 17, 21, 19, 13, 13 and 17. What are their Mode and Range? _____

Score: ....

Answer Key

1) Mode: 8, Range: 11	2) Mode: 13, Range: 17
3) Mode: 7, Range: 6	4) Mode: 12, Range: 6
5) Mode: 2, Range: 8	6) Mode: 1, Range: 19
7) Mode: 7, Range: 30	8) Mode: 6, Range: 11
9) Mode: 22, Range: 32	10) Mode: 13, Range: 11

Times Series

- A precise representation of the distribution of numerical data is referred as Time Series.

EXAMPLE:

Use the following Graph to complete the table.

Day	Distance (km)
1	
2	

→

Answer:

Day	Distance (km)
1	359
2	460
3	278
4	547
5	360

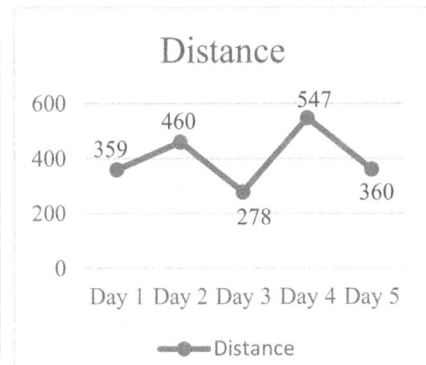

PRACTICES:

Use the following Graph to complete the table.

1)

Day	Distance (km)
1	
2	

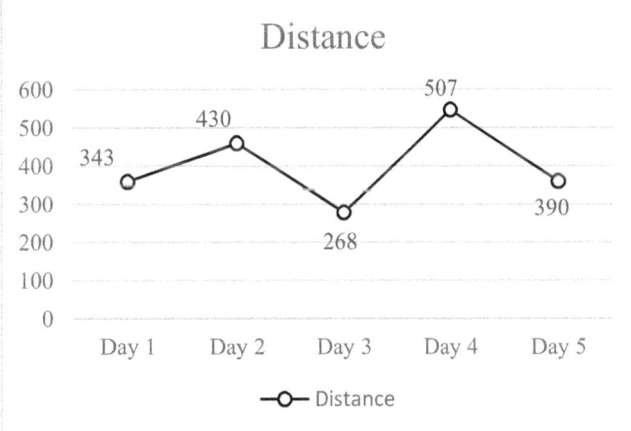

The following table shows the number of births in the US from 2007 to 2012 (in millions).

2) Draw a time series for the table.

Year	Number of births (in millions)
2007	6.42
2008	6.45
2009	6.33
2010	5.9
2011	4.35
2012	4.35

Score: ....

Answer Key

1)

Day	Distance (km)
1	343
2	430
3	268
4	507
5	390

2)

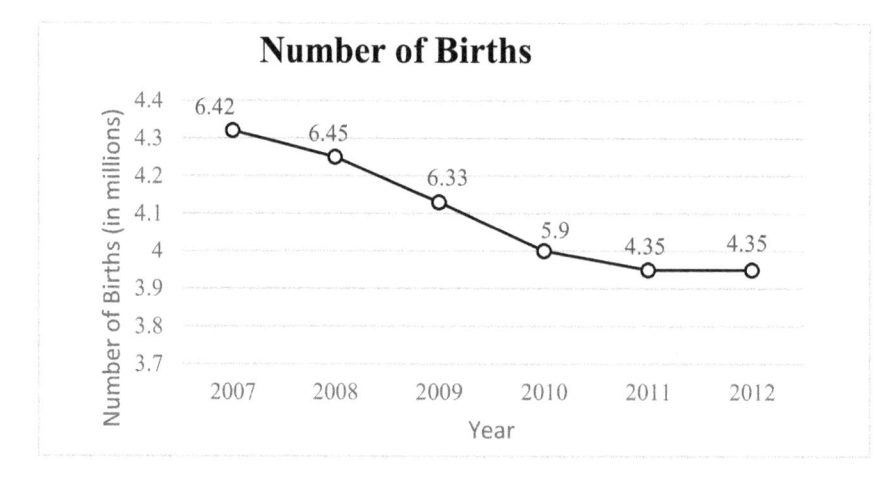

Box and Whisker Plot

- ✓ Box-and-whisker plots display data including quartiles.
- ✓ IQR – interquartile range shows the difference from Q1 to Q3.
- ✓ Extreme Values are the lowest and highest values in a data set.

EXAMPLE:

73, 84, 86, 95, 68, 67, 100, 94, 77, 80, 62, 79
Maximum: 100, Minimum: 62; Q_1: 70.5; Q_2: 79.5; Q_3: 90

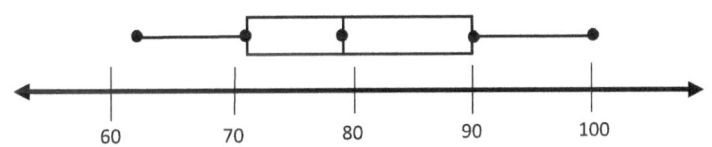

PRACTICES:

Make box and whisker plots for the given data.

1) 1, 5, 20, 8, 3, 10, 13, 11, 14, 17, 18, 15, 23

2) 2, 7, 23, 11, 13, 9, 16, 5, 18, 22, 20, 17, 19

3) 3, 7, 9, 10, 11, 5, 14, 19, 20, 21, 22, 8, 14

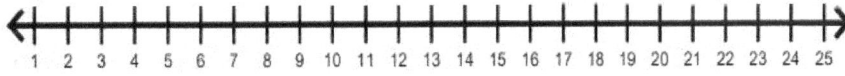

4) 4, 6, 5, 15, 12, 14, 10, 7, 21, 17, 8, 22, 6

Answer Key

1) 1, 3, 5, 8, 10, 11, 13, 14, 15, 17, 18, 20, 23

   Maximum: 23, Minimum: 1, Q_1: 8, Q_2: 13, Q_3: 17

   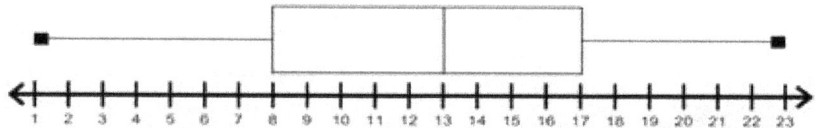

2) 2, 7, 23, 11, 13, 9, 16, 5, 18, 22, 20, 17, 19

   Maximum: 23, Minimum: 2, Q_1: 9, Q_2: 16, Q_3: 19

   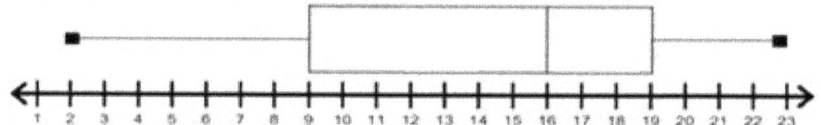

3) 3, 7, 9, 10, 11, 5, 14, 19, 20, 21, 22, 8, 14

   Maximum: 22, Minimum: 2, Q_1: 8, Q_2: 11, Q_3: 19

   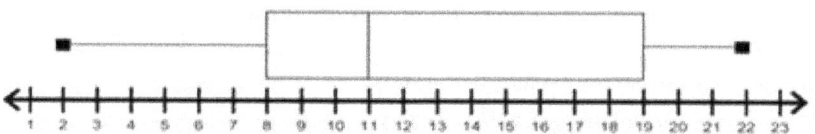

4) 4, 6, 5, 15, 12, 14, 10, 7, 21, 17, 8, 22, 6

   Maximum: 22, Minimum: 4, Q_1: 6, Q_2: 10, Q_3: 15

   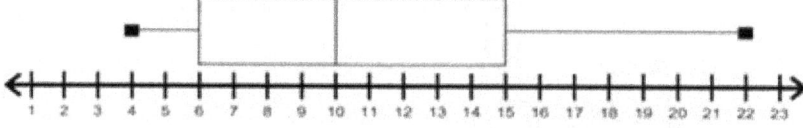

Bar Graph

✓ A chart that presents data with bars in different heights to match with the values of the data is called a bar graph. We can graph the bars vertically or horizontally.

EXAMPLE:

Graph the given information as a bar graph.

Name of the Sport	Total Number of Students
Football	15
Volleyball	7
Table Tennis	7
Basketball	12
Chess	9

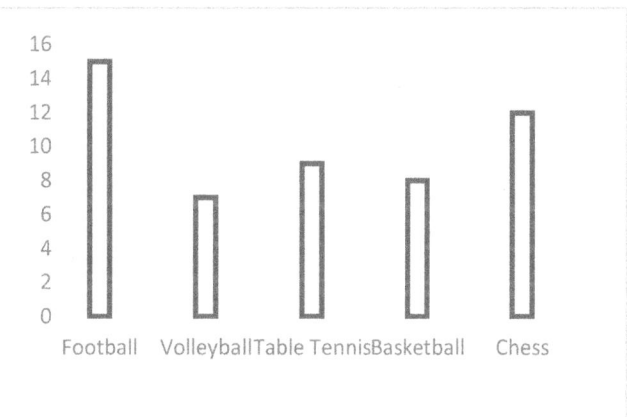

PRACTICES:

Graph the given information as a bar graph.

1)

Day	Sale House
Monday	6
Tuesday	4
Wednesday	10
Thursday	5
Friday	2
Saturday	8
Sunday	1

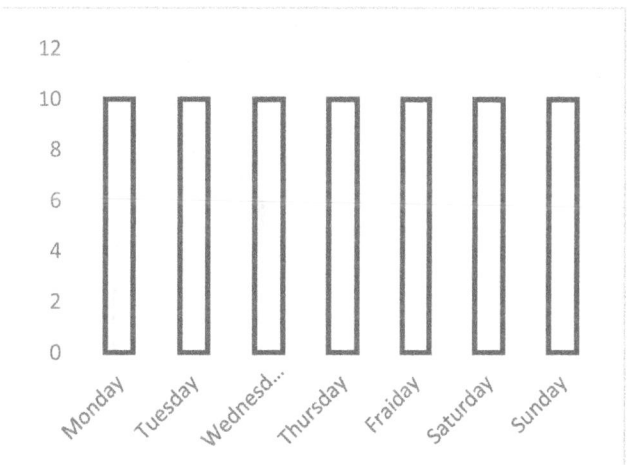

TASC Math Prep

2)

Day	Sale House
Monday	8
Tuesday	6
Thursday	3
Friday	10
Saturday	4
Sunday	2

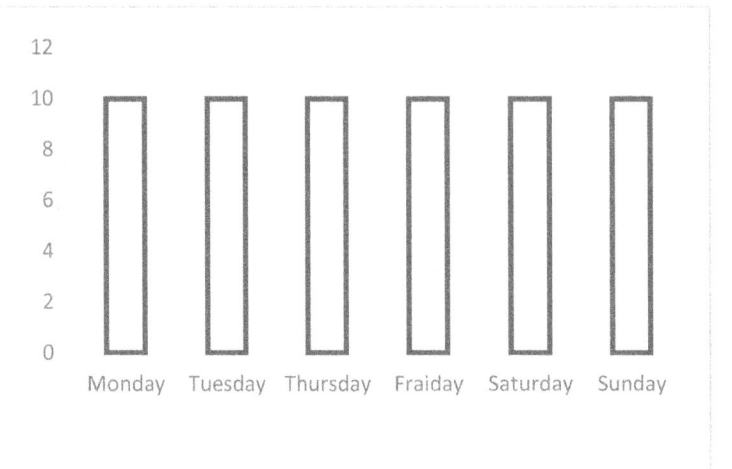

Score: ....

Answer Key

1)

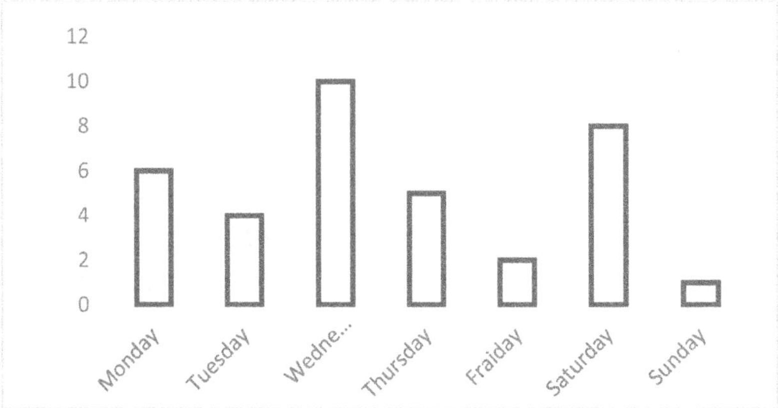

2)

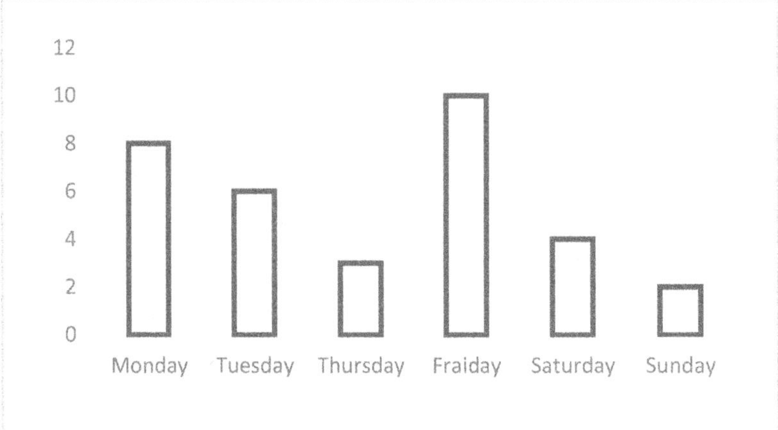

Dot plots

- ✓ The representation of a distribution that consists of group of data points plotted on a simple scale is referred as a dot plot. Dot plots are used for continuous, univariate and quantitative data. If there are few data points, then they can be labelled.
- ✓ Dot plots are one of the simplest statistical plots and are suitable for small to moderate sized data sets.

EXAMPLE:

A survey of "How many books each student purchased?" has these results How many students purchase 4 books?

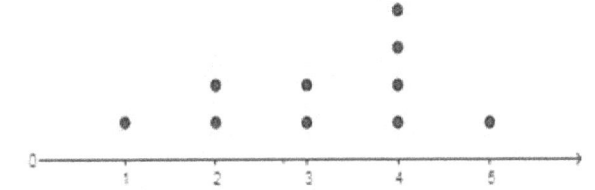

4 students

PRACTICES:

A survey of "How many pets each person owned?" has these results:

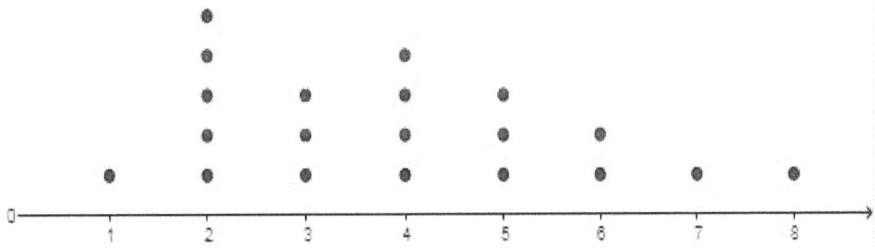

1) How many people have at least 3 pets?

2) How many people have 2 and 3 pets?

3) How many people have 4 pets?

4) How many people have 2 or less than 2 pets?

5) How many people have more than 7 pets?

6) How many people have more than 4 pets?

Score: ....

Answer Key

1) 4	2) 8
3) 4	4) 6
5) 1	6) 7

Scatter Plots

- ✓ The values with points that represent the relationship between two sets of data are shown by a scatter (x, y) plot.
- ✓ The horizontal values are taken as x and vertical data is taken as y.

EXAMPLE:

Construct a scatter plot.

x	1	2	3	4	5
y	2	4.5	1.5	5	2

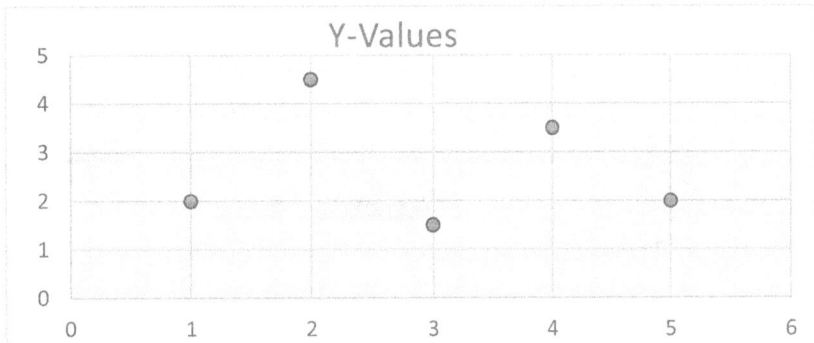

PRACTICES:

Construct a scatter plot.

1)

x	1	2.5	3	3.5	4	5
y	4	3.5	4.5	2.5	8	2

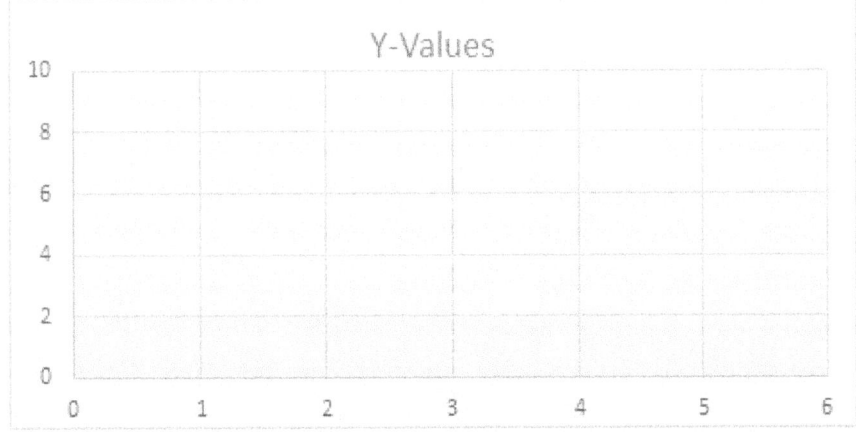

TASC Math Prep

2)

x	1	2	3.5	4	4.5	5
y	3	1	2.5	1.5	1.5	1

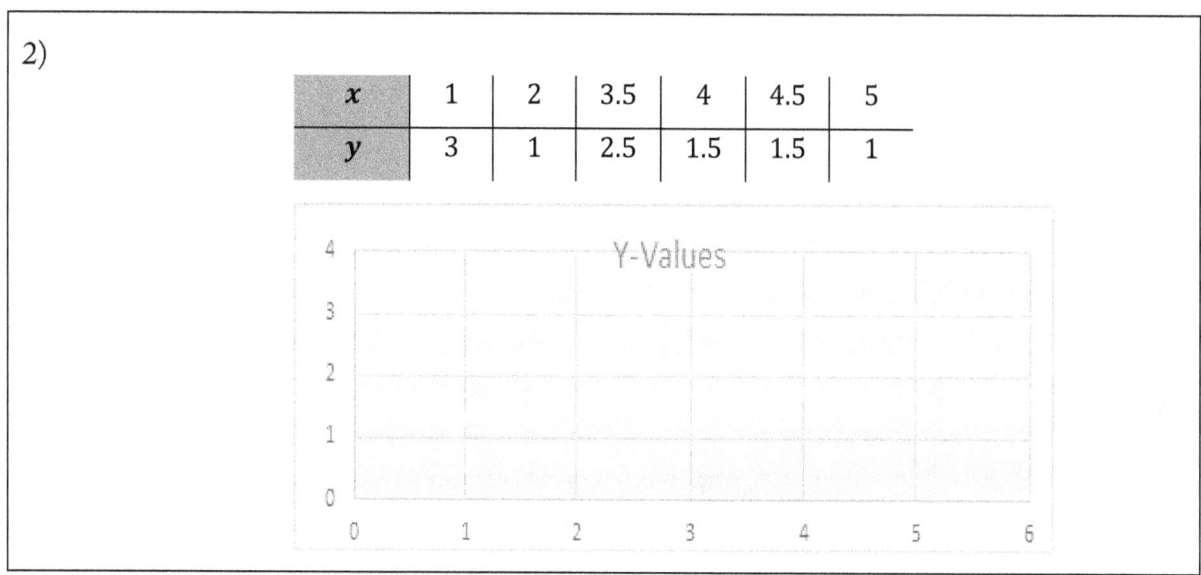

Score: ....

Answer Key

1)

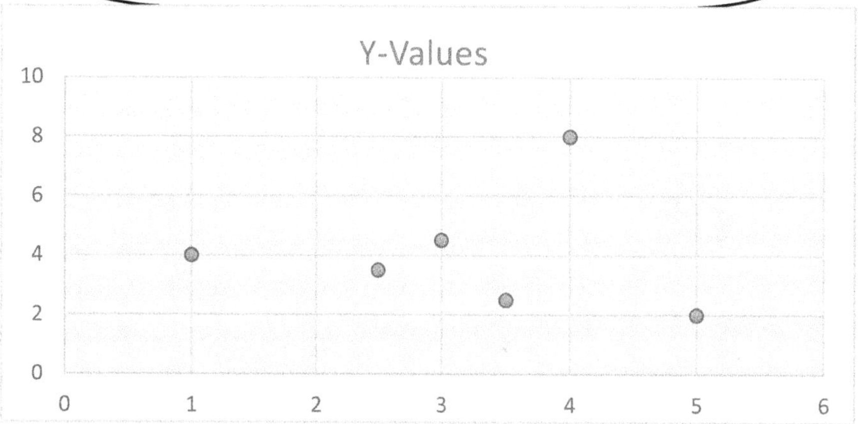

2)

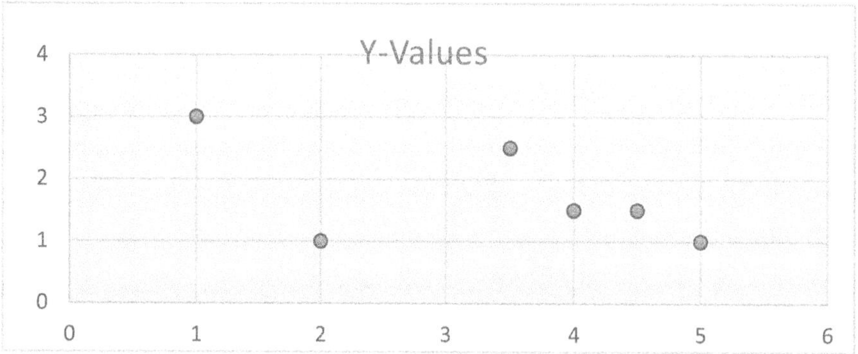

Stem–And–Leaf Plot

- ✓ Stem-and-leaf plots display the frequency of the values in a data set.
- ✓ We can make a frequency distribution table for the values, or we can use a stem-and-leaf plot.

EXAMPLE:

56, 58, 42, 48, 66, 64, 53, 69, 45, 72

Stem	leaf
4	2 5 8
5	3 6 8
6	4 6 9
7	2

PRACTICES:

Make stem ad leaf plots for the given data.

1) 42, 14, 17, 21, 44, 24, 18, 47, 23, 24, 19, 12

Stem | Leaf plot

2) 10, 65, 14, 18, 69, 11, 33, 61, 66, 38, 15, 35

Stem | Leaf plot

3) 122, 87, 99, 86, 100, 126, 92, 129, 88, 121, 91, 107

Stem | Leaf plot

4) 60, 51, 119, 69, 72, 59, 110, 65, 77, 59, 65, 112, 71

Stem | Leaf plot

Score: ....

Answer Key

1)

Stem	leaf
1	2 4 7 8 9
2	1 3 4
4	2 4 7

2)

Stem	leaf
1	0 1 4 5 8
3	3 5 8
6	1 5 6 9

3)

Stem	leaf
8	6 7 8
9	1 2 9
10	0 7
12	1 2 6 9

4)

Stem	leaf
5	1 9 9
6	0 5 5 9
7	1 2 7
11	0 2 9

The Pie Graph or Circle Graph

✓ A circular chart divided into sectors is said to be a pie chart; each sector represents the relative size of each value.

EXAMPLE:

A library has 840 books that include Mathematics, Physics, Chemistry, English, and History. Use following graph to answer question.

What is the number of Mathematics books?

Number of total books = 840,

Percent of Mathematics books = 30% = 0.30

Then: 0.30 × 840 = 252

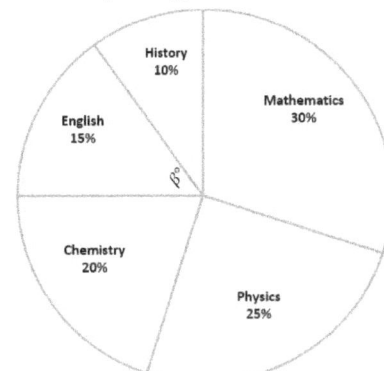

PRACTICES:

Favorite Sports:

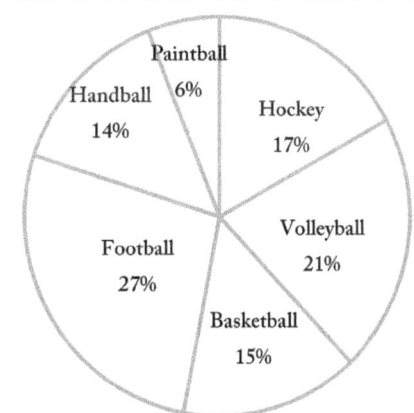

SPORTS

1) What percentage of pie graph is paintball?
2) What percentage of pie graph is Hockey and volleyball?
3) What percentage of pie not Football and Basketball?
4) What percentage of pie graph is Hockey and Handball and Football?
5) What percentage of pie graph is Basketball?
6) What percentage of pie not Handball and Paintball?

Score: ....

Answer Key

1) 6%	2) 38%
3) 58%	4) 58%
5) 15%	6) 80%

Name: ....

Probability of Simple Events

- ✓ Probability is the possibility of something happening in the future. It is shown as a number between zero (can never happen) to 1 (will always happen).
- ✓ Probability can be written as a fraction, a decimal, or a percent.

EXAMPLE:

If there are 8 red balls and 12 blue balls in a basket, what is the probability that John will pick out a red ball from the basket?

There are 8 red ball and 20 are total number of balls. Therefore, probability that John will pick out a red ball from the basket is 8 out of 20 or $\frac{8}{8+12} = \frac{8}{20} = \frac{2}{5}$

PRACTICES:

Solve.

1) A number is chosen at random from 28 to 35. Find the probability of selecting factors of 5.

2) A number is chosen at random from 1 to 60. Find the probability of selecting multiples of 15.

3) Find the probability of selecting 4queens from a deck of card.

4) A number is chosen at random from 8 to 19. Find the probability of selecting factors of 3.

5) What probability of selecting a ball less than 6 from 10 different bingo balls?

6) A number is chosen at random from 1 to 10. What is the probability of selecting a multiple of 2?

7) A card is chosen from a well-shuffled deck of 52 cards. What is the probability that the card will be a king OR a queen?

8) A number is chosen at random from 1 to 20. What is the probability of selecting multiples of 5?

9) A number is chosen at random from 1 to 10. Find the probability of selecting number 4 or smaller numbers.

10) Bag A contains 9 red marbles and 3 green marbles. Bag B contains 9 black marbles and 6 orange marbles. What is the probability of selecting a green marble at random from bag A? What is the probability of selecting a black marble at random from Bag B?

Score: ....

Answer Key

1) $\frac{1}{4}$	2) $\frac{1}{15}$
3) $\frac{1}{13}$	4) $\frac{1}{3}$
5) $\frac{1}{2}$	6) $\frac{1}{2}$
7) $\frac{2}{13}$	8) $\frac{1}{5}$
9) $\frac{2}{5}$	10) $\frac{1}{4}, \frac{3}{5}$

Chapter 11 : TASC Test Review

The Test Assessing Secondary Completion, commonly known as the TASC or high school equivalency degree, is a standardized test. The TASC is a standardized test to verify that examinees have knowledge in core content areas equivalent to that of graduating high school seniors.

There are five subject area tests on TASC:
- Reading
- Writing
- Social Studies
- Science
- Mathematics.

The TASC Mathematics test is a 105-minute test that covers basic mathematics topics, quantitative problem-solving and algebraic questions. There are two Mathematics sections on the TASC. The first section contains 40 multiple choice questions where calculators are permitted. You have 50 minutes to complete this section. The second section contains 12 Gridded-Response questions. Calculator is NOT allowed in the second part. Test takers have 55 minutes to answer all questions in this section. Examinees will also be given a page of mathematic formulas to use during the test.

In this book, we have covered all Mathematics topics you need to know. Now, it's time to take a real TASC test. In this section, there are two complete TASC Mathematics Tests. Take these tests to see what score you'll be able to receive on a real TASC test.

Time to Test

Time to refine your quantitative reasoning skill with a practice test

In this section, there are two complete TASCT Mathematics practice tests. Take these tests to simulate the test day experience. After you've finished, score your tests using the answer keys.

Before You Start

- You'll need a pencil, a calculator and a timer to take the test.
- For each question, there are four possible answers. Choose which one is best.
- It's okay to guess. There is no penalty for wrong answers.
- Use the answer sheet provided to record your answers.
- After you've finished the test, review the answer key to see where you went wrong.

Good Luck!

Mathematics is like love; a simple idea, but it can get complicated.

TASC Practice Tests Answer Sheet

Remove (or photocopy) these answer sheets and use them to complete the practice tests.

TASC Practice Test – Section 1 Answer Sheet

1	Ⓐ Ⓑ Ⓒ Ⓓ	9	Ⓐ Ⓑ Ⓒ Ⓓ	17	Ⓐ Ⓑ Ⓒ Ⓓ	25	Ⓐ Ⓑ Ⓒ Ⓓ	33	Ⓐ Ⓑ Ⓒ Ⓓ
2	Ⓐ Ⓑ Ⓒ Ⓓ	10	Ⓐ Ⓑ Ⓒ Ⓓ	18	Ⓐ Ⓑ Ⓒ Ⓓ	26	Ⓐ Ⓑ Ⓒ Ⓓ	34	Ⓐ Ⓑ Ⓒ Ⓓ
3	Ⓐ Ⓑ Ⓒ Ⓓ	11	Ⓐ Ⓑ Ⓒ Ⓓ	19	Ⓐ Ⓑ Ⓒ Ⓓ	27	Ⓐ Ⓑ Ⓒ Ⓓ	35	Ⓐ Ⓑ Ⓒ Ⓓ
4	Ⓐ Ⓑ Ⓒ Ⓓ	12	Ⓐ Ⓑ Ⓒ Ⓓ	20	Ⓐ Ⓑ Ⓒ Ⓓ	28	Ⓐ Ⓑ Ⓒ Ⓓ	36	Ⓐ Ⓑ Ⓒ Ⓓ
5	Ⓐ Ⓑ Ⓒ Ⓓ	13	Ⓐ Ⓑ Ⓒ Ⓓ	21	Ⓐ Ⓑ Ⓒ Ⓓ	29	Ⓐ Ⓑ Ⓒ Ⓓ	37	Ⓐ Ⓑ Ⓒ Ⓓ
6	Ⓐ Ⓑ Ⓒ Ⓓ	14	Ⓐ Ⓑ Ⓒ Ⓓ	22	Ⓐ Ⓑ Ⓒ Ⓓ	30	Ⓐ Ⓑ Ⓒ Ⓓ	38	Ⓐ Ⓑ Ⓒ Ⓓ
7	Ⓐ Ⓑ Ⓒ Ⓓ	15	Ⓐ Ⓑ Ⓒ Ⓓ	23	Ⓐ Ⓑ Ⓒ Ⓓ	31	Ⓐ Ⓑ Ⓒ Ⓓ	39	Ⓐ Ⓑ Ⓒ Ⓓ
8	Ⓐ Ⓑ Ⓒ Ⓓ	16	Ⓐ Ⓑ Ⓒ Ⓓ	24	Ⓐ Ⓑ Ⓒ Ⓓ	32	Ⓐ Ⓑ Ⓒ Ⓓ	40	Ⓐ Ⓑ Ⓒ Ⓓ

TASC Practice Test: Section 2: Grid-ins Questions

41 [grid]   42 [grid]   43 [grid]   44 [grid]

45
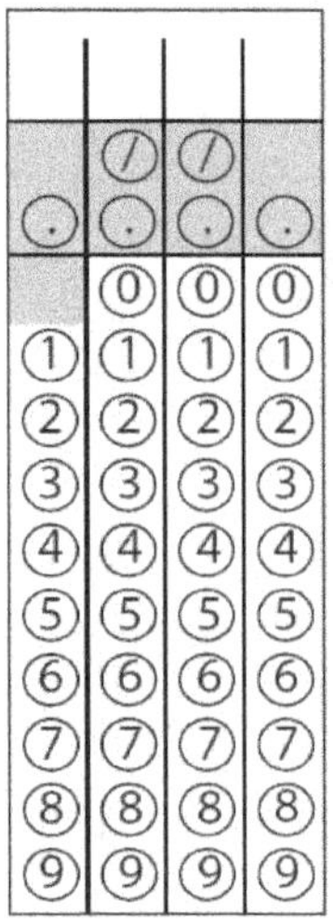

46
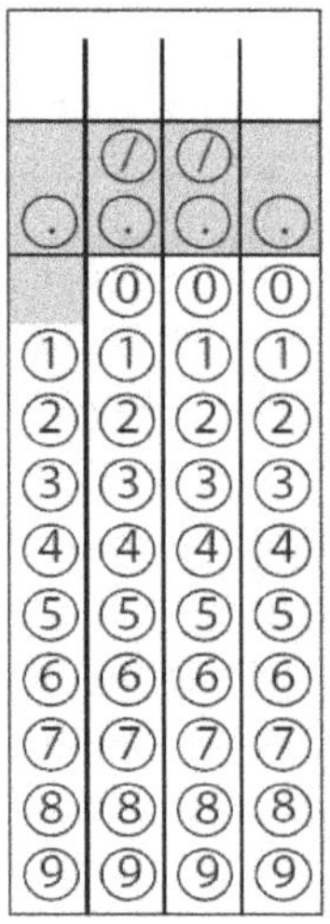

47
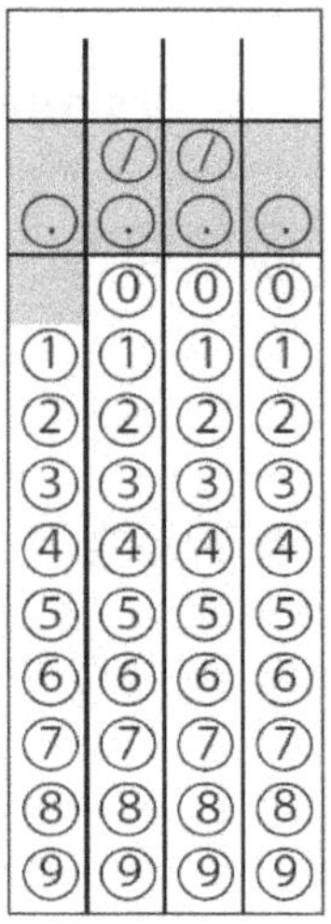

48
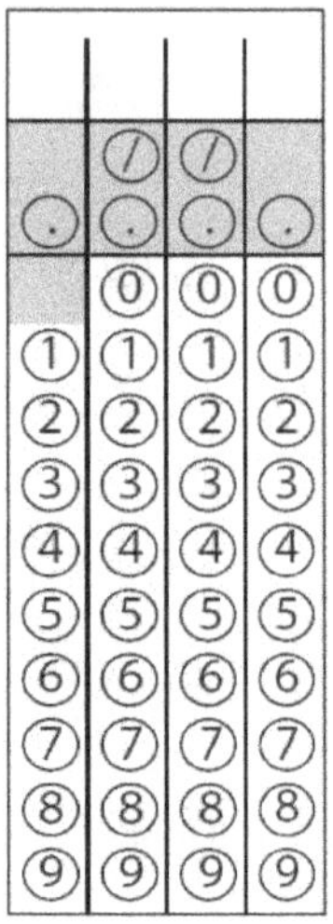

49
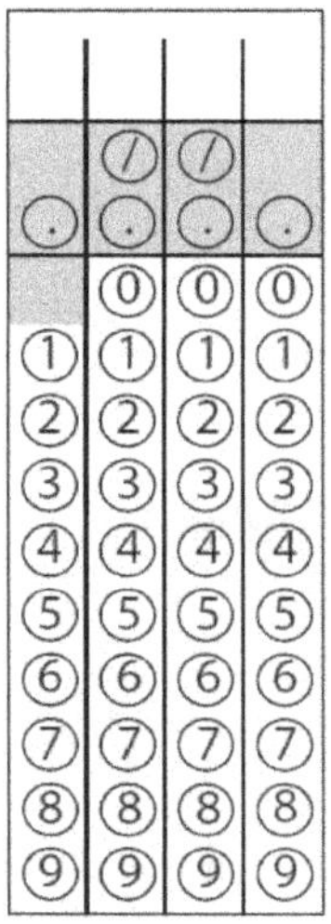

50
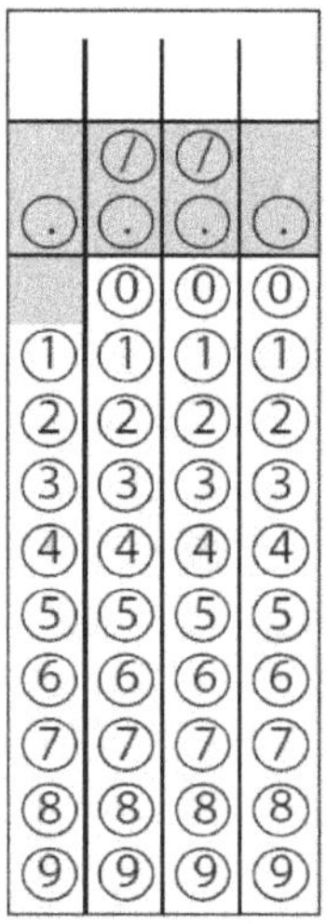

51
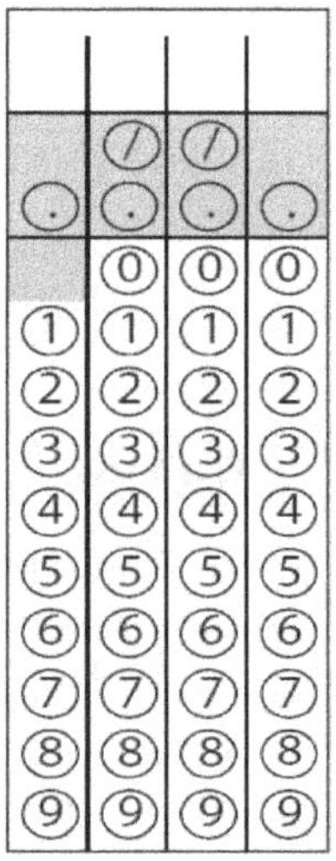

52
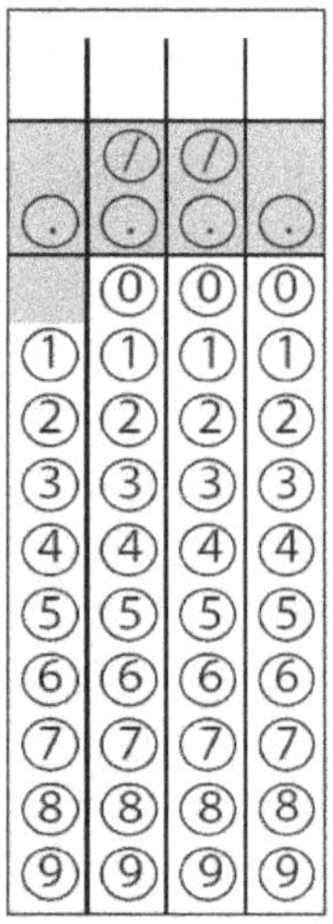

TASC Mathematics Reference Sheet

Volume

Cylinder: $v = \pi r^2 h$

Pyramid: $v = \frac{1}{3} bh$

cone: $v = \frac{1}{3} \pi r^2 h$

Sphere: $\frac{4}{3} \pi r^3$

Coordinate Geometry

Midpoint of the segment AB:

$M\left(\frac{x_1+x_2}{2}, \frac{y_1+y_2}{2}\right)$

Distance from A to B:

$d = \sqrt{(x_1 - x_2)^2 + (y_1 - y_2)^2}$

Slope of a line:

$m = \frac{y_2 - y_1}{x_2 - x_1} = \frac{rise}{run}$

Special Factoring

$a^2 - b^2 = (a + b)(a - b)$

$a^2 + 2ab + b^2 = (a + b)(a + b)$

$a^2 - 2ab + b^2 = (a - b)(a - b)$

$a^3 + b^3 = (a + b)(a^2 - ab + b^2)$

$a^3 - b^3 = (a - b)(a^2 + ab + b^2)$

Quadratic Formula

for $ax^2 + bx + c = 0$

$x = \frac{-b \pm \sqrt{b^2 - 4ac}}{2a}$

Interest

Simple Interest:

$I = prt$

Interest Formula (compounded n times per year):

$A = p\left(1 + \frac{r}{n}\right)^{nt}$

A = Amount after t years.

p = principal

r = annual interest rate

t = time in years

I = Interest

Trigonometric Identities

Pythagorean Theorem:

$a^2 + b^2 = c^2$

$\sin \theta = \frac{opp}{hyp}$

$\cos \theta = \frac{adj}{hyp}$

$\tan \theta = \frac{opp}{adj}$

$\sin^2 \theta + \cos^2 \theta = 1$

Density = $\frac{Mass}{Volume}$

Central Angle	Inscribed Angle	Intersecting Chords Theorem
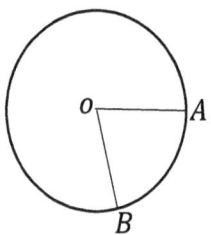 $m\angle AOB = m\widehat{AB}$	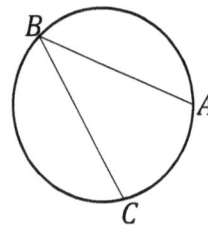 $m\angle ABC = \frac{1}{2} m\widehat{AC}$	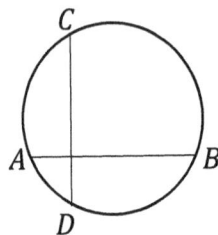 $A \cdot B = C \cdot D$

Probability

Permutations: $_nP_r = \frac{n!}{(n-r)!}$

Combinations: $_nC_r = \frac{n!}{(n-r)!\,r!}$

Multiplication rule (independent events): $P(A \text{ and } B) = P(A) \cdot P(B)$

Multiplication rule (general): $P(A \text{ and } B) = P(A) \cdot P(B|A)$

Addition rule: $P(A \text{ or } B) = P(A) + P(B) - P(A \text{ and } B)$

Conditional Probability: $P(B|A) = \frac{P(A \text{ and } B)}{P(A)}$

Arithmetic Sequence: $a_n = a_1 + (n-1)d$ where a_n is the nth term, a_1 is the first term, and d is the common difference.

Geometric Sequence: $a_n = a_1 r^{(n-1)}$ where a_n is the nth term, a_1 is the first term, and r is the common ratio.

TASC Practice Test 1

Mathematics

Section 1 - Calculator

- ✓ 40 questions
- ✓ Total time for this section: 50 Minutes
- ✓ You may use a calculator on this Section.

Administered *Month Year*

1) 46 is What percent of 40?

   A. 115 %

   B. 64 %

   C. 36 %

   D. 136 %

2) The marked price of a computer is E Euro. Its price decreased by 25% in March and later increased by 8 % in April. What is the final price of the computer in E Euro?

   A. 0.79 E

   B. 0.91 E

   C. 0.81 E

   D. 1.06 E

3) A rope weighs 350 grams per meter of length. What is the weight in kilograms of 12.4 meters of this rope? (1 kilograms = 1000 grams)

   A. 0.4034

   B. 4.34

   C. 4.3

   D. 43.30

4) Which of the following could be the product of two consecutive prime numbers? (Select one or more answer choices)

   A. 120

   B. 77

   C. 144

   D. 26

5) A $50 shirt now selling for $15 is discounted by what percent?

   A. 65 %

   B. 21 %

   C. 54 %

   D. 70 %

TASC Math Prep

6) The score of Zoe was one-fifth of Emma and the score of Harper was triple that of Emma. If the score of Harper was 105, what is the score of Zoe?

   A. 7

   B. 35

   C. 24

   D. 14

7) Three fourth of 32 is equal to $\frac{3}{5}$ of what number?

   A. 20

   B. 30

   C. 40

   D. 60

8) A bag contains 25 balls: nine green, four black, four blue, seven red and one white. If 24 balls are removed from the bag at random, what is the probability that a white ball has been removed?

   A. $\frac{1}{24}$

   B. $\frac{23}{24}$

   C. $\frac{1}{25}$

   D. $\frac{24}{25}$

9) What is the value of 5^4?

   A. 196

   B. 625

   C. 343

   D. 125

10) What is the median of these numbers? 14, 7, 18, 30, 10, 10, 9, 14, 26

    A. 7

    B. 26

    C. 9

    D. 14

11) How many tiles of 2 cm² is needed to cover a floor of dimension 4 cm by 14 cm?

   A. 28   C. 52

   B. 24.5   D. 26

12) Ryan traveled 280 km in 8 hours and Riley traveled 252 km in 6 hours. What is the ratio of the average speed of Ryan to average speed of Riley?

   A. 7: 5   C. 7: 6

   B. 6: 5   D. 5: 6

13) An angle is equal to one fifth of its supplement. What is the measure of that angle?

   A. 30   C. 15

   B. 18   D. 20

14) Abigail purchased a sofa for $386.88 The sofa is regularly priced at $624. What was the percent discount Abigail received on the sofa?

   A. 22 %   C. 38 %

   B. 62 %   D. 24 %

15) Find the average of the following numbers: 21, 16, 24 and 12?

   A. 20.5   C. 16

   B. 18.25   D. 18

16) When a number is subtracted from 45 and the difference is divided by that number, the result is 4. What is the value of the number?

   A. 13   C. 7

   B. 9   D. 10

17) Right triangle ABC has two legs of lengths 8 cm (AB) and 6 cm (AC). What is the length of the third side (BC)?

　　A. 3 cm　　　　　　　　　　C. 10 cm

　　B. 6 cm　　　　　　　　　　D. 12 cm

18) A taxi driver earns $16 per hour work. If he works 12 hours a day, and he uses 3-liters Petrol in 2 hours with price $2.50 for 1-liter. How much money does he earn in one day?

　　A. $147　　　　　　　　　　C. $18

　　B. $160　　　　　　　　　　D. $45

19) The width of a box is one fourth of its length. The height of the box is one half of its width. If the length of the box is 48 cm, what is the volume of the box?

　　A. 645 cm^3　　　　　　　　C. 524 cm^3

　　B. 768 cm^3　　　　　　　　D. 3,456 c m

20) The price of a sofa is decreased by 25% to $465. What was its original price?

　　A. $610　　　　　　　　　　C. $600

　　B. $620　　　　　　　　　　D. $360

21) If 40 % of a class are girls, and 35 % of girls play tennis, what percent of the class play tennis?

　　A. 24 %　　　　　　　　　　C. 14 %

　　B. 26 %　　　　　　　　　　D. 39 %

22) The price of a car was $20,000 in 2014, $16,000 in 2015 and $12,800 in 2016. What is the rate of depreciation of the price of car per year?

A. 25 %

B. 30 %

C. 20 %

D. 35 %

23) The average of three consecutive numbers is 43. What is the smallest number?

A. 45

B. 43

C. 44

D. 42

24) The area of a circle is less than 64π. Which of the following can be the circumference of the circle? (Select one or more answer choices)

A. 24π

B. 49π

C. 12π

D. 16π

25) A bank is offering 2.25% simple interest on a savings account. If you deposit $8,000, how much interest will you earn in four years?

A. $135

B. $4,200

C. $720

D. $12,500

26) In four successive hours, a car travels 40 km, 46 km, 43 km, and 450km. In the next four hours, it travels with an average speed of 45 km per hour. Find the total distance the car traveled in 8 hours.

A. 354 km

B. 360 km

C. 352 km

D. 704 km

27) The ratio of boys to girls in a school is 4:3. If there are 420 students in a school, how many boys are in the school.

   A. 140

   B. 120

   C. 240

   D. 90

28) If the area of trapezoid is 168 cm^2, what is the perimeter of the trapezoid?

   A. 10

   B. 86

   C. 58

   D. 60

29) In the xy-plane, the point $(3, -4)$ and $(5, 4)$ are on the line A. Which of the following points could also be on the line A?

   A. $(1, 4)$

   B. $(2, 2)$

   C. $(1, -12)$

   D. $(0, 3)$

30) A chemical solution contains 8% alcohol. If there is 19.2 ml of alcohol, what is the volume of the solution?

   A. 420 ml

   B. 360 ml

   C. 480 ml

   D. 240 ml

31) The average high of 15 constructions in a town is 140 m, and the average high of 5towers in the same town is 160 m. What is the average high of all the 20 structures in that town?

   A. 160

   B. 145

   C. 150

   D. 140

32) The price of a laptop is decreased by 15% to $425. What is its original price?

   A. 450

   B. 488.75

   C. 500

   D. 361.25

33) In 1999, the average worker's income increased $1,800 per year starting from $24,500 annual salary. Which equation represents income greater than average? (I = income, x = number of years after 1999)

   A. $I > 1,800\,x + 24,500$

   B. $I > -1,800\,x + 24,500$

   C. $I < 24,500\,x + 1,800$

   D. $I < -24,500\,x + 1,800$

34) A boat sails 160 miles south and then 120 miles east. How far is the boat from its start point?

   A. 90 miles

   B. 160 miles

   C. 120 miles

   D. 200 miles

35) If 85 % of F is 17 % of M, then F is what percent of M?

   A. 7 %

   B. 70 %

   C. 80 %

   D. 500 %

36) How many possible outfit combinations come from nine shirts, seven slacks, and 5 times?

   A. 20

   B. 315

   C. 20!

   D. 315!

37) The surface area of a cylinder is $36\pi\ cm^2$. If its height is 7 cm, what is the radius of the cylinder?

   A. 9 cm

   B. 3cm

   C. 2 cm

   D. 4 cm

38) How long does a 351–miles trip take moving at 45 miles per hour (mph)?

   A. 8 hours

   B. 7 hours and 48 minutes

   C. 8 hours and 48 minutes

   D. 7 hours and 36 minutes

39) A shirt costing $500 is discounted 20%. After a month, the shirt is discounted another 10%. Which of the following expressions can be used to find the selling price of the shirt?

   A. $(500) - 500(0.30)$

   B. $(500)(0.80)(0.90)$

   C. $(500)(0.20) - (340)(0.10)$

   D. $(500)(0.20) - (500)(0.10)$

40) Which of the following lists shows the fractions in order from least to greatest?

   A. $\frac{3}{5}, \frac{8}{11}, \frac{7}{9}$

   B. $\frac{8}{11}, \frac{3}{5}, \frac{7}{9}$

   C. $\frac{7}{9}, \frac{3}{5}, \frac{8}{11}$

   D. $\frac{3}{5}, \frac{7}{9}, \frac{8}{11}$

IF YOU FINISH BEFORE TIME IS CALLED, YOU MAY CHECK YOUR WORK ON THIS SECTION ONLY. DO NOT TURN TO OTHER SECTION IN THE TEST.

STOP

TASC Practice Test 1

Mathematics

Section 2 - No Calculator

- ✓ 12 questions
- ✓ Total time for this section: 55 Minutes
- ✓ You may NOT use a calculator on this Section.

Administered *Month Year*

41) From last year, the price of gasoline has increased from $1.50 per gallon to $1.86 per gallon. The new price is what percent of the original price?

42) If $4x - 5 = 2$, What is the value of $3x - \frac{3}{4}$?

43) What is the area of an isosceles right triangle that has one leg that measures 10 cm?

44) What is the value of $f(3)$ for the following function f?

$$f(x) = 3x^2 - 4x - 13$$

45) If the ratio of $2a$ to $7b$ is $\frac{1}{10}$, what is the ratio of a to b?

46) A construction company is building a wall. The company can build 16 cm of the wall per minute. After 50 minutes $\frac{8}{9}$ of the wall is completed. How many meters is the wall?

47) $8 \times (-5) + 22 - 3(-5 - 16 \times 5) \div 15 = ?$

48) The perimeter of the trapezoid below is 58 cm. What is its area?

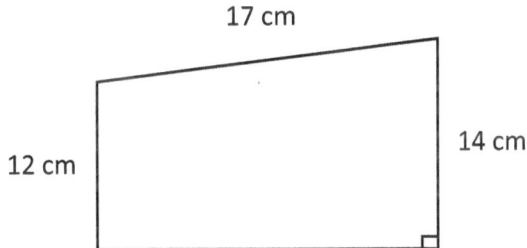

49) The sum of two numbers is M. if one of the numbers is 5, then twice of the other number would be what?

50) The volume of cube A is $\frac{1}{3}$ of its surface area. What is the length of an edge of cube A?

51) If $\frac{3x-6}{4} = N$ and $N = -3$, what is the value of x?

52) A ladder leans against a wall forming a 45° angle between the ground and the ladder. If the bottom of the ladder is 16 feet away from the wall, how many feet is the ladder?

TASC Practice Test 2

Mathematics

Section 1 - Calculator

- ✓ 40 questions
- ✓ Total time for this section: 50 Minutes
- ✓ You may use a calculator on this Section.

Administered *Month Year*

1) What is the perimeter of a square in centimeters that has an area of 691.69 cm²?

   A. 24.4

   B. 105.2

   C. 95.5

   D. 148.8

2) If 75% of A is 25% of B, then B is what percent of A?

   A. 33%

   B. 66%

   C. 3%

   D. 300%

3) Simplify the expression.

$$(7x^3 - 4x^2 - 4x^4) - (6x^2 - 2x^4 + 5x^3)$$

   A. $-3(2x^4 + 2x^3 - 5x^2)$

   B. $2(x^4 + x^3 - 2x^2)$

   C. $-2(x^4 - x^3 + 5x^2)$

   D. $(-3x^4 - 3x^3 - 7x^2)$

4) Mr. Matthews saves $4,200 out of his annually family income of $71,400. What fractional part of his income does he save?

   A. $\frac{1}{17}$

   B. $\frac{12}{17}$

   C. $\frac{15}{17}$

   D. $\frac{16}{17}$

5) What is the median of these numbers? 26, 10, 5, 35, 29, 18, 52

   A. 52

   B. 12.5

   C. 35

   D. 26

6) A bank is offering 2.05% simple interest on a savings account. If you deposit $14,000, how much interest will you earn in five years?

   A. $1,000

   B. $1,435

   C. $2,280

   D. $12,400

7) Last week 14,000 fans attended a football match. This week four times as many bought tickets, but one seventh of them cancelled their tickets. How many are attending this week?

   A. 84,000

   B. 62,000

   C. 8,000

   D. 48,000

8) In two successive years, the population of a town is increased by 7% and 25%. What percent of the population is increased after two years?

   A. 16.2%

   B. 33.8%

   C. 12.2 %

   D. 25 %

9) The mean of 50 test scores was calculated as 66. But it turned out that one of the scores was misread as 58 but it was 38. What is the correct mean of the test scores?

   A. 66

   B. 65

   C. 65.5

   D. 66.5

10) Which of the following graphs represents the compound inequality $-4 \leq 3x - 7 < 11$?

   A.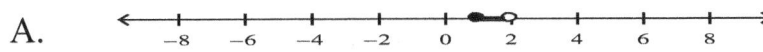

   B.

   C.

   D.

11) Two dice are thrown simultaneously, what is the probability of getting a sum of 4 or 9?

A. $\frac{3}{36}$

B. $\frac{7}{18}$

C. $\frac{1}{9}$

D. $\frac{7}{36}$

12) Which of the following shows the numbers in descending order?

$$\frac{1}{5}, 0.8, 9\%, \frac{1}{13}$$

A. $\frac{1}{5}, 0.9, \frac{1}{13}, 0.8$

B. $9\%, \frac{1}{5}, 0.8, \frac{1}{13}$

C. $\frac{1}{13}, 9\%, \frac{1}{5}, 0.8$

D. $0.8, \frac{1}{5}, 9\%, \frac{1}{13}$

13) What is the area of a square whose diagonal is 8?

A. 18

B. 32

C. 24

D. 36

14) What is the volume of a box with the following dimensions?

Hight = 4 cm   Width = 9 cm   Length = 5 cm

A. 180 cm³

B. 59 cm³

C. 36 cm³

D. 64 cm³

15) The average of 8 numbers is 32. The average of 5 of those numbers is 50. What is the average of the other three numbers?

A. 6

B. 12

C. 2

D. 32

16) What is the value of x in the following system of equations?

$$2x + y = -5$$
$$3x - 2y = 10$$

A. 4

B. -4

C. 5

D. -5

17) The perimeter of a rectangular yard is 72 meters. What is its length if its width is triple its length?

A. 9 meters

B. 24 meters

C. 12 meters

D. 32 meters

18) In a stadium the ratio of home fans to visiting fans in a crowd is 8:4. Which of the following could be the total number of fans in the stadium? (Select one or more answer choices)

A. 43,600

B. 39,300

C. 51,100

D. 50,600

19) What is the equivalent temperature of $104°F$ in Celsius? $C = \frac{5}{9}(F - 32)$

A. 42

B. 32

C. 40

D. 54

20) Which of the following points lies on the line $5x - 3y = 1$? (Select one or more answer choices)

A. $(-1, 0)$

B. $(3, -1)$

C. $(-1, -2)$

D. $(2, 1)$

TASC Math Prep

21) Mr. Jefferson family are choosing a menu for their reception. They have 4 choices of appetizers, 7 choices of entrees, 6 choices of cake. How many different menu combinations are possible for them to choose?

   A. 15

   B. 72

   C. 85

   D. 168

22) Anita's trick–or–treat bag contains 21 pieces of chocolate, 16 suckers, 13 pieces of gum, 15 pieces of licorice. If she randomly pulls a piece of candy from her bag, what is the probability of her pulling out a piece of gum?

   A. $\frac{1}{52}$

   B. $\frac{1}{13}$

   C. $\frac{1}{65}$

   D. $\frac{1}{5}$

23) The average of five numbers is 42. If a sixth number that is greater than 48 is added, then, which of the following could be the new average? (Select one or more answer choices)

   A. 44

   B. 38

   C. 39

   D. 42

24) The ratio of boys and girls in a class is 5:7. If there are 60 students in the class, how many more boys should be enrolled to make the ratio 1:1?

   A. 10

   B. 7

   C. 12

   D. 25

25) The perimeter of the trapezoid below is 32 cm. What is its area?

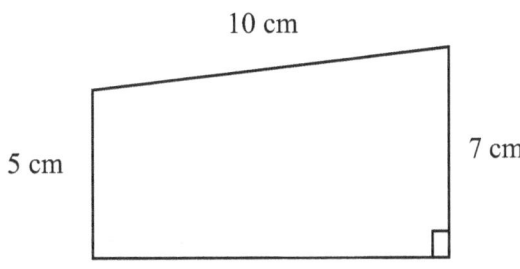

A. 180 cm²

B. 60 cm²

C. 140 cm²

D. 168 cm²

26) A card is drawn at random from a standard 52–card deck, what is the probability that the card is of hearts? (The deck includes 13 of each suit clubs, diamonds, hearts, and spades)

A. $\frac{4}{52}$

B. $\frac{1}{4}$

C. $\frac{1}{52}$

D. $\frac{1}{13}$

27) What is the surface area of the cylinder below?

A. 150 π in²

B. 354 π in²

C. 244 π in²

D. 200 π in²

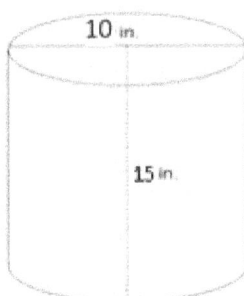

28) Simplify $5x^2y^5(-3xy^2)^3$.

A. $5x^4y^7$

B. $-27x^5y^{12}$

C. $-135x^5y^{11}$

D. $135x^5y^{11}$

29) Daniel is 20 miles ahead of Noa and running at 4.5 miles per hour. Noa is running at the speed of 7 miles per hour. How long does it take Noa to catch Daniel?

A. 1 hour, and 40 minutes

B. 8 hours, 30 minutes

C. 8 hours

D. 9 hours

30) A football team had $35,000 to spend on supplies. The team spent $16,000 on new balls. New sport shoes cost $135 each. Which of the following inequalities represent the number of new shoes the team can purchase?

A. $135x + 19,000 \leq 35,000$

B. $135x + 19,000 \leq 35,000$

C. $19,000 + 135x \geq 35,000$

D. $19,000 + 135x \geq 35,000$

31) If 60 % of a number is 21, what is the number?

A. 6

B. 20

C. 9

D. 35

32) The square of a number is $\frac{147}{192}$. What is the cube of that number?

A. $\frac{54}{74}$

B. $\frac{343}{512}$

C. $\frac{36}{49}$

D. $\frac{1,125}{2,305}$

33) 70 students took an exam and 21 of them failed. What percent of the students passed the exam?

A. 80 %

B. 20 %

C. 70 %

D. 30 %

34) The circle graph below shows all Mr. Wilson's expenses for last month. If he spent $630 on his car, how much did he spend for his rent?

A. $840

B. $235.20

C. $1,400

D. $1,210

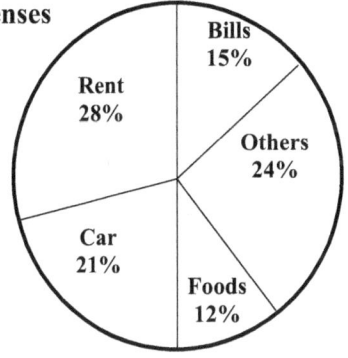

Mr. Wilson's monthly expenses

35) What is the value of x in the following equation?

$$\frac{5}{9}x + \frac{1}{6} = \frac{1}{2}$$

A. 5

B. $\frac{3}{8}$

C. $\frac{3}{5}$

D. $\frac{5}{3}$

36) If 130 % of a number is 104, then what is the 80 % of that number?

A. 44

B. 73

C. 64

D. 63

37) Mrs. Thomson needs an 85% average in her writing class to pass. On her first 3 exams, he earned scores of 72%, 92%, and 84%. What is the minimum score Mrs. Thomson can earn on her fourth and final test to pass?

A. 92

B. 89

C. 90.5

D. 87.5

38) The length of a rectangle is 4 meters less than 7 times its width. The perimeter of the rectangle is 72 meters. What is the area of the rectangle in meters?

   A. 115

   B. 136

   C. 155

   D. 36

39) A swimming pool holds 3,600 cubic feet of water. The swimming pool is 24 feet long and 15 feet wide. How deep is the swimming pool?

   A. 1.6

   B. 10

   C. 75

   D. 120

40) Right triangle ABC has two legs of lengths 12 cm (AB) and 16 cm (AC). What is the length of the third side (BC)?

   A. 4 cm

   B. 6 cm

   C. 8 cm

   D. 20 cm

TASC Practice Test 2

Mathematics

Section 2 - No Calculator

- ✓ 12 questions
- ✓ Total time for this section: 55 Minutes
- ✓ You may NOT use a calculator on this Section.

Administered *Month Year*

41) A tree 80 feet tall casts a shadow 12 feet long. Jack is 5 feet tall. How long is Jack's shadow?

42) What is the value of the expression $-2(5x + y) + (1 - 5x)^2$ when $x = 1.2$ and $y = -7$?

43) What is the slope of a line that is perpendicular to the line $x - 4y = 8$?

44) $[7 \times (-8) - 42] + [(-3) \times (-14)] \div 7 - (-14) = ?$

45) What is the product of all possible values of x in the following equation?

$$|-2x + 1| = 17$$

46) What is the value of x in the following equation? $-37 = 86 - x$

47) If $4x - 7 = 3$, what is the value of $3x + \frac{3}{2}$?

48) The width of a box is one fourth of its length. The height of the box is one fourth of its width. If the length of the box is 32 cm, what is the volume of the box?

49) The area of a rectangular yard is 210 square meters. What is its width if its length is 15 meters?

50) In a classroom of 50 students, 20 are female. What percentage of the class is male?

51) One fifth of 90 is equal to $\frac{3}{7}$ of what number?

52) The average weight of 25 girls in a class is 35 kg and the average weight of 5 boys in the same class is 38 kg. What is the average weight of all the 32 students in that class?

Chapter 12: Answers and Explanations

Answer Key

❋Now, it's time to review your results to see where you went wrong and what areas you need to improve!

TASC Mathematics Practice Tests

TASC Practice Test 1

	Section 1				Section 2	
1	A	16	B	31	B	41	124
2	C	17	C	32	C	42	4.5
3	B	18	A	33	A	43	50
4	B	19	D	34	D	44	2
5	D	20	B	35	D	45	0.35
6	A	21	C	36	B	46	9
7	C	22	C	37	C	47	-1
8	D	23	D	38	B	48	195
9	B	24	C	39	B	49	$2M-10$
10	D	25	C	40	A	50	2
11	A	26	A			51	-2
12	D	27	C			52	$16\sqrt{2}$
13	A	28	D				
14	C	29	C				
15	B	30	D				

TASC Practice Test 2

	Section 1				Section 2	
1	B	16	D	31	D	41	0.75
2	D	17	A	32	B	42	27
3	C	18	B	33	A	43	-4
4	A	19	C	34	C	44	-78
5	D	20	C	35	C	45	-72
6	B	21	D	36	C	46	123
7	D	22	D	37	A	47	9
8	B	23	A	38	C	48	512
9	C	24	B	39	B	49	14
10	D	25	A	40	D	50	60
11	D	26	B			51	42
12	C	27	D			52	53.25
13	B	28	C				
14	A	29	C				
15	C	30	B				

Practice Test 1
Section 1- Calculator

1) Answer: A

Use percent formula: $Part = \frac{percent \times whole}{100}$

$46 = \frac{percent \times 40}{100} \Rightarrow \frac{46}{1} = \frac{percent \times 40}{100}$, cross multiply.

$4,600 = percent \times 40$, divide both sides by 40: $115 = percent$

2) Answer: C

To find the discount, multiply the number by (100% – rate of discount).

Therefore, for the first discount we get: $(100\% - 25\%)(E) = (0.75)E$

For increase of 8 %: $(0.75)E \times (100\% + 8\%) = (0.75)(1.08) = 0.81E$.

3) Answer: B

The weight of 12.4 meters of this rope is: $12.4 \times 350g = 4,340g$

1 kg = 1,000 g, therefore, $4,340\ g \div 1000 = 4.34 kg$

4) Answer: B

Some of prime numbers are: 2, 3, 5, 7, 11, 13

Find the product of two consecutive prime numbers:

$7 \times 11 = 77$ (bingo!)

$11 \times 13 = 143$ (not in the options)

$13 \times 17 = 221$ (not in the options)

Choice B is correct.

5) Answer: D

Use the formula for Percent of Change $= \frac{New\ Value - Old\ Value}{Old\ Value} \times 100\%$

$\frac{50-15}{50} \times 100\% = -70\%$ (negative sign here means that the new price is less than old price).

6) Answer: A

If the score of Harper was 105, therefore the score of Emma is 35. Since, the score of Zoe was one fifth of Emma, therefore, the score of Zoe is 7.

7) Answer: C

Let x be the number. Write the equation and solve for x.

$\frac{3}{4} \times 32 = \frac{3}{5} \cdot x \Rightarrow \frac{3 \times 32}{4} = \frac{3x}{5}$, use cross multiplication to solve for x.

$15 \times 32 = 3x \times 4 \Rightarrow 480 = 12x \Rightarrow x = 40$

8) Answer: D

If 24 balls are removed from the bag at random, there will be one ball in the bag.

The probability of choosing a white ball is 1 out of 25. Therefore, the probability of not choosing a white ball is 24 out of 25 and the probability of having not a white ball after removing 24 balls is the same.

9) Answer: B

$5^4 = 5 \times 5 \times 5 \times 5 = 625$

10) Answer: D

Write the numbers in order:

7, 9, 10, 10, 14, 14, 18, 26, 30

Since we have 9 numbers (9 is odd), then the median is the number in the middle, which is 14.

11) Answer: A

The area of the floor is: $4\ cm \times 14\ cm = 56\ cm^2$

The number of tiles needed $= 56 \div 2 = 28$

12) Answer: D

The average speed of Ryan is: $280 \div 8 = 35$ km

The average speed of Riley is: $252 \div 6 = 42$ km

Write the ratio and simplify. $35 : 42 \Rightarrow 5 : 6$.

13) Answer: A

The sum of supplement angles is 180. Let x be that angle. Therefore,

$x + 5x = 180$.

$6x = 180$, divide both sides by 6: $x = 30$.

14) Answer: C

Use percent formula: $Part = \frac{percent \times whole}{100}$

$386.88 = \frac{percent \times 624}{100} \Rightarrow$ (cross multiply): $38{,}688 = percent \times 624 \Rightarrow$

$percent = \frac{38{,}688}{624} = 62$

386.88 is 62 % of 624. Therefore, the discount is: 100% – 62% = 38%.

15) Answer: B

$average = \frac{sum\ of\ terms}{number\ of\ terms} \Rightarrow average = \frac{(21+16+24+12)}{4} \Rightarrow \frac{73}{4} = 18.25$

16) Answer: B

Let x be the number. Write the equation and solve for x.

$\frac{(45-x)}{x} = 4$ (cross multiply)

$(45 - x) = 4x$, then add x both sides. $45 = 5x$, now divide both sides by 5. $\Rightarrow x = 9$.

17) Answer: C

Use Pythagorean Theorem: $a^2 + b^2 = c^2$

$8^2 + 6^2 = C^2 \Rightarrow 64 + 36 = C^2 \Rightarrow 100 = c^2 \Rightarrow c = 10$

18) Answer: A

$16 \times 12 = \$192$

Petrol use: $3 \div 2 = 1.5$, $12 \times 1.5 = 18$ liters

Petrol cost: $18 \times \$2.50 = \$45$

Money earned: $\$192 - \$45 = \$147$

19) Answer: D

If the length of the box is 48, then the width of the box is one fourth of it, 12, and the height of the box is 6 (one half of the width). The volume of the box is:

$V = lwh = (48)(12)(6) = 3{,}456$

20) Answer: B

Let x be the original price.

If the price of the sofa is decreased by 25% to $465, then: 75 % of $x = 465 \Rightarrow 0.75x = 465 \Rightarrow x = 465 \div 0.75 = 620$

TASC Math Prep

21) Answer: C

The percent of girls playing tennis is: $40\% \times 35\% = 0.40 \times 0.35 = 0.14 = 14\%$

22) Answer: C

Use this formula: Percent of Change

$$\frac{\text{New Value} - \text{Old Value}}{\text{Old Value}} \times 100\%$$

$\frac{16,000 - 20,000}{20,000} \times 100\% = -20\%$ and $\frac{12,800 - 16,000}{16,000} \times 100\% = -20\%$

23) Answer: D

Let x be the smallest number. Then, these are the numbers:

$x, x+1, x+2$

$average = \frac{\text{sum of terms}}{\text{number of terms}} \Rightarrow 43 = \frac{x + (x+1) + (x+2)}{3} \Rightarrow 43 = \frac{3x+3}{3} \Rightarrow 43 = x+1 \Rightarrow x = 42$

24) Answer: C

Area of the circle is less than 64π. Use the formula of areas of circles.

$$Area = \pi r^2 \Rightarrow \pi r^2 < 64\pi \Rightarrow r^2 < 64 \Rightarrow r < 8$$

Radius of the circle is less than 8. Let's put 8 for the radius. Now, use the circumference formula:

$$Circumference = 2\pi r = 2\pi(8) = 16\pi$$

Since the radius of the circle is less than 7. Then, the circumference of the circle must be less than 16π. Choices C is less than 16π

25) Answer: C

Use simple interest formula: $I = prt$ (I= interest, p = principal, r = rate, t = time)

$$I = (8,000)(0.0225)(4) = 720$$

26) Answer: A

Add the first 6 numbers. $40 + 46 + 43 + 45 = 174$

To find the distance traveled in the next 5 hours, multiply the average by number of hours.

$Distance = Average \times Rate = 45 \times 4 = 180$

Add both numbers. $174 + 180 = 354$

27) Answer: C

The ratio of boy to girls is 4: 3. Therefore, there are 4 boys out of 7 students. To find the answer, first divide the total number of students by 7, then multiply the result by 4.

$420 \div 7 = 60 \Rightarrow 60 \times 4 = 240$

28) Answer: D

The area of the trapezoid is: $Area = \frac{1}{2} h(b_1 + b_2) = \frac{1}{2}(x)(24 + 18) = 168$

$\rightarrow 21x = 168 \rightarrow x = 8$

$y = \sqrt{6^2 + 8^2} = \sqrt{36 + 64} = \sqrt{100} = 10$

The perimeter of the trapezoid is: $8 + 10 + 18 + 24 = 60$

29) Answer: C

The equation of a line is in the form of $y = mx + b$, where m is the slope of the line and b is the $y - intercept$ of the line.

Two points $(3, -4)$ and $(5, 4)$ are on the line A. Therefore, the slope of the line A is:

$slope\ of\ line\ A = \frac{y_2 - y_1}{x_2 - x_1} = \frac{4-(-4)}{5-3} = \frac{8}{2} = 4$

The slope of line A is 4. Thus, the formula of the line A is:

$y = mx + b = 4x + b$, choose a point and plug in the values of x and y in the equation to solve for b. Let's choose point $(5, 4)$. Then:

$$y = 4x + b \rightarrow 4 = 20 + b \rightarrow b = 4 - 20 = -16$$

The equation of line A is: $y = 4x - 16$

Now, let's review the choices provided:

A. $(1, 4)$ $y = 4x - 16 \rightarrow 4 = 4 - 16 = -12$,   This is not true.

B. $(2, 2)$ $y = 4x - 16 \rightarrow 2 = 8 - 16 = -8$,   This is not true.

C. $(1, -12)$ $y = 4x - 16 \rightarrow -12 = 4 - 16 = -12$,   This is true!

D. $(0, 3)$ $y = 4x - 16 \rightarrow 3 = 0 - 16 = -16$,   This is not true.

30) Answer: D

8% of the volume of the solution is alcohol. Let x be the volume of the solution. Then:

8% of x = 19.2 ml

$0.08\ x = 19.2 \Rightarrow \frac{8x}{100} = \frac{192}{10}$ cross multiply

$80x = 19,200 \Rightarrow$ (devide by 80) $x = 240$

31) **Answer: B**

$average = \frac{sum\ of\ terms}{number\ of\ terms}$

The sum of the high of all constructions is: $15 \times 140 = 2,100$ m

The sum of the high of all towers is: $5 \times 160 = 800$ m

The sum of the high of all building is: $2,100 + 800 = 2,900$

$average = \frac{2,900}{20} = 145$

32) **Answer: C**

Let x be the original price.

If the price of a laptop is decreased by 15% to $425, then:

$85\ \%\ of\ x = 425 \Rightarrow 0.85x = 425 \Rightarrow x = 425 \div 0.85 = 500$

33) **Answer: A**

Let x be the number of years. Therefore, $1,800 per year equals $1,800x$.

starting from $24,500 annual salary means you should add that amount to $1,800x$.

Income more than that is: I > $1,800x + 24,500$

34) **Answer: D**

Use the information provided in the question to draw the shape.

Use Pythagorean Theorem: $a^2 + b^2 = c^2 \Rightarrow 120^2 + 160^2 = c^2$

$\Rightarrow 14,400 + 25,600 = c^2 \Rightarrow 40,000 = c^2 \Rightarrow c = 200$.

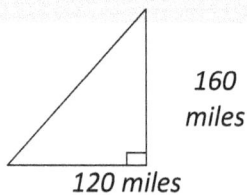

35) **Answer: D**

Write the equation and solve for M:

$0.85\ F = 0.17\ M$, divide both sides by 0.17, then: $\frac{0.85}{0.17} F = M$, therefore:

$M = 5\ F$, and M is 5 times of F or it's 500% of F.

36) **Answer: B**

To find the number of possible outfit combinations, multiply number of options for each factor: 9 × 7 × 5 = 315

37) Answer: C

Formula for the Surface area of a cylinder is:

$SA = 2\pi r^2 + 2\pi rh \rightarrow 36\pi = 2\pi r^2 + 2\pi r(7) \rightarrow r^2 + 7r - 18 = 0$

$(r + 9)(r - 2) = 0 \rightarrow r = 2 \quad or \quad r = -9 \; (unacceptable)$

38) Answer: B

Use distance formula:

$Distance = Rate \times time \Rightarrow 351 = 45 \times T$, divide both sides by 45. $\Rightarrow T = 7.8$ hours.

Change hours to minutes for the decimal part. $0.8 \; hours = 0.8 \times 60 = 48 \; minutes$.

39) Answer: B

To find the discount, multiply the number by (100% – rate of discount).

Therefore, for the first discount we get: (500) (100% – 20%) = (500) (0.8)

For the next 10 % discount: (500) (0.80) (0.90).

40) Answer: A

Let's compare each fraction:

$\frac{3}{5} < \frac{8}{11} < \frac{7}{9}$

Only choice A provides the right order.

TASC Practice Test 1
Section 2 - No Calculator

41) Answer: 124

The question is this: 1.86 is what percent of 1.50?

Use percent formula:

$$part = \frac{percent}{100} \times whole$$

$1.86 = \frac{percent}{100} \times 1.50 \Rightarrow 1.86 = \frac{percent \times 1.50}{100} \Rightarrow 186 = percent \times 1.50 \Rightarrow percent = \frac{186}{1.50} = 124.$

42) Answer: 4.5

$4x - 5 = 2 \rightarrow 4x = 2 + 5 = 7 \rightarrow x = \frac{7}{4} = 1.75$

Then; $3x - \frac{3}{4} = 3(1.75) - \frac{3}{4} = 5.25 - 0.75 = 4.5$.

43) Answer: 50

First draw an isosceles triangle. Remember that two sides of the triangle are equal. Let put a for the legs. Then:

$a = 10 \Rightarrow$ area of the triangle is $= \frac{1}{2}(10 \times 10) = \frac{100}{2} = 50\ cm^2$.

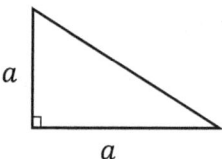

Isosceles right triangle

44) Answer: 2

The input value is 3. Then: $x = 3$

$f(x) = 3x^2 - 4x - 13 \rightarrow f(3) = 3(3^2) - 4(3) - 13 = 27 - 12 - 13 = 2$

45) Answer: 0.35

Write the ratio of $2a$ to $7b$.

$\frac{2a}{7b} = \frac{1}{10}$

Use reciprocal of $\frac{2}{7}$ and multiply both sides, then simplify.

$\frac{2a}{7b} = \frac{1}{10} \Rightarrow \frac{7}{2} \times \frac{2a}{7b} = \frac{1}{10} \times \frac{7}{2} \Rightarrow \frac{a}{b} = \frac{7}{20} = 0.35$

46) Answer: 9 m

The rate of construction company = $\frac{16 \text{ cm}}{1 \text{ min}}$ = 16 cm/min

Height of the wall after 50 minutes = $\frac{16 \text{ cm}}{1 \text{ min}} \times 50$ min = 800 cm

Let x be the height of wall, then $\frac{8}{9}x = 800$ cm → $x = \frac{9 \times 800}{8}$ → $x = 900$ cm = $9\ m$

47) Answer: −1

Use PEMDAS (order of operation):

$8 \times (-5) + 22 - 3(-5 - 16 \times 5) \div 15 = -40 + 22 - 3(-5 - 80) \div 15 = -18 - 3(-85) \div 15 = -18 + 255 \div 15 = -18 + 17 = -1$.

48) Answer: 195.

The perimeter of the trapezoid is 58.

Therefore, the missing side (height) is $= 58 - (12 + 14 + 17) = 15$ cm

Area of a trapezoid: $A = \frac{1}{2} h (b1 + b2) = \frac{1}{2}(15)(12 + 14) = 195\ cm^2$

49) Answer: $2M - 10$

$x + y = M$ and $x = 5 \Rightarrow 5 + y = M \Rightarrow y = M - 5, \Rightarrow 2y = 2(M - 5) = 2M - 10$

50) Answer: 2

Let x be the length of an edge of cube, then the volume of a cube is: $V = x^3$

The surface area of cube is: $SA = 6x^2$

The volume of cube A is $\frac{1}{3}$ of its surface area. Then:

$x^3 = \frac{6x^2}{3} \to x^3 = 2x^2$, divide both side of the equation by x^2. Then:

$$\frac{x^3}{x^2} = \frac{2x^2}{x^2} \to x = 2$$

51) Answer: −2

Since $N = -3$, substitute -3 for N in the equation $\frac{3x-6}{4} = N$, which gives $\frac{3x-6}{4} = -3$.

Multiplying both sides by 4 gives $3x - 6 = -12$ and then adding 6 to both sides, then $3x = -6$ then, Dividing both sides by 3 then, $x = -2$.

52) Answer: $16\sqrt{2}$

The relationship among all sides of special right triangle 45°; 45°; 90° is provided in this triangle:

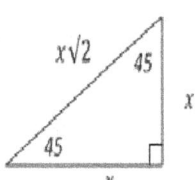

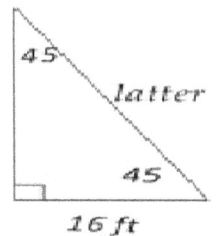

In this triangle, the opposite side of 45° angle is half of the hypotenuse.

Draw the shape of this question.

The latter is the hypotenuse. Therefore, the latter is $16\sqrt{2}$ feet.

Practice Test 2
Section 1 - Calculator

1) Answer: B

The area of the square is 691.69. Therefore, the side of the square is square root of the area.

$\sqrt{691.69} = 26.3$

Four times the side of the square is the perimeter: $4 \times 26.3 = 105.2$

2) Answer: D

Write the equation and solve for B: 0.75 A = 0.25 B, divide both sides by 0.25, then:
$\frac{0.75}{0.25}$ A = B, therefore: B = 3 A, and B is 3 times of A or it's 300% of A.

3) Answer: C

Simplify and combine like terms.

$(7x^3 - 4x^2 - 4x^4) - (6x^2 - 2x^4 + 5x^3)$

$\Rightarrow (7x^3 - 4x^2 - 4x^4) - 6x^2 + 2x^4 - 5x^3$

$\Rightarrow -2x^4 + 2x^3 - 10x^2 = -2(x^4 - x^3 + 5x^2)$.

4) Answer: A

4,200 out of 71,400 equals to $\frac{4,200}{71,400} = \frac{42}{714} = \frac{7}{119} = \frac{1}{17}$

5) Answer: D

Write the numbers in order: 5, 10, 18, 26, 29, 35, 52

Median is the number in the middle. So, the median is 26.

6) Answer: B

Use simple interest formula: $I = prt$ (I = interest, p = principal, r = rate, t = time)

$I = (14,000)(0.0205)(5) = 1,435$

7) Answer: D

Four times of 14,000 is 56,000. One seventh of them cancelled their tickets.

One seventh of 56,000 equal 8,000 ($\frac{1}{7} \times 56,000 = 8,000$).

$(56,000 - 8,000 = 48,000)$ fans are attending this week

8) Answer: B

the population is increased by 7% and 25%. 7% increase changes the population to 107% of original population.

For the second increase, multiply the result by 125%.

$(1.07) \times (1.25) = 1.338 = 133.8\%$

33.8 percent of the population is increased after two years.

9) Answer: C

average (mean) = $\frac{\text{sum of terms}}{\text{number of terms}} \Rightarrow 66 = \frac{\text{sum of terms}}{50} \Rightarrow$ sum = $66 \times 50 = 3,300$

The difference of 58 and 38 is 20. Therefore, 20 should be subtracted from the sum.

$3,300 - 20 = 3,280$

mean = $\frac{\text{sum of terms}}{\text{number of terms}} \Rightarrow$ mean = $\frac{3,280}{50} = 65.5$

10) Answer: D

Solve for x.

$-4 \leq 3x - 7 < 11 \Rightarrow$ (add 7 all sides) $-4 + 7 \leq 3x - 7 + 7 < 7 + 11 \Rightarrow 3 \leq 3x < 18$

$\Rightarrow$ (divide all sides by 3) $1 \leq x < 6$

x is between 1 and 6. Choice C represent this inequality.

11) Answer: D

To get a sum of 4 for two dice, we can get 3 different options:

$(1, 3), (2, 2), (3, 1)$

To get a sum of 9 for two dice, we can get 4 different options:

$(3, 6), (6, 3), (4, 5), (5, 4)$

Therefore, there are 7 options to get the sum of 5 or 7.

Since, we have $6 \times 6 = 36$ total options, the probability of getting a sum of 5 or 7 is 7 out of 36 or $\frac{7}{36}$.

12) Answer: C

Change the numbers to decimal and then compare.

$\frac{1}{5} = 0.2 \ldots$

0.8

9% = 0.09

$\dfrac{1}{13} = 0.07$

Therefore $\dfrac{1}{13} < 9\% < \dfrac{1}{5} < 0.8$.

13) **Answer: B**

The diagonal of the square is 8. Let x be the side.

Use Pythagorean Theorem: $a^2 + b^2 = c^2$

$x^2 + x^2 = 8^2 \Rightarrow 2x^2 = 64 \Rightarrow x^2 = 32 \Rightarrow x = \sqrt{32}$

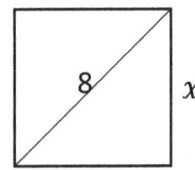

The area of the square is: $\sqrt{32} \times \sqrt{32} = 32$

14) **Answer: A**

Volume of a box = length × width × height = $4 \times 9 \times 5 = 180$

15) **Answer: C**

average = $\dfrac{\text{sum of terms}}{\text{number of terms}} \Rightarrow 32 = \dfrac{\text{sum of 8 numbers}}{8} \Rightarrow$ sum of 8 numbers = $32 \times 8 = 256$

$50 = \dfrac{\text{sum of 5 numbers}}{5} \Rightarrow$ sum of 5 numbers = $5 \times 50 = 250$

sum of 8 numbers − sum of 5 numbers = sum of 3 numbers

$256 - 250 = 6 \rightarrow$ average of 3 numbers $= \dfrac{6}{3} = 2$.

16) **Answer: D**

Solving Systems of Equations by Elimination

$2x + y = -5$
$3x - 2y = 10$ Multiply the first equation by 3, and second equation by −2, then add two equations.

$\begin{array}{l} 3(2x + y = -5) \\ -2(3x - 2y = 10) \end{array} \Rightarrow \begin{array}{l} 6x + 3y = -15 \\ -6x + 4y = -20 \end{array} \Rightarrow 7y = -35 \Rightarrow y = -5.$

17) **Answer: A**

The width of the rectangle is triple its length. Let x be the length. Then, $width = 3x$

Perimeter of the rectangle is 2 (width + length) = $2(3x + x) = 72 \Rightarrow 8x = 72 \Rightarrow x = 9$

Length of the rectangle is 9 meters.

TASC Math Prep

18) Answer: B

In the stadium the ratio of home fans to visiting fans in a crowd is 8:4. Therefore, total number of fans must be divisible by 12: $8 + 4 = 12$.

Let's review the choices:

- A. 43,600: $43,600 \div 12 = 3,633.3$
- B. 39,300: $39,300 \div 12 = 3,275$
- C. 51,100: $51,100 \div 12 = 4,258.3$
- D. 50,600: $50,600 \div 12 = 4,216.6$

Only choice B when divided by 12 result a whole number.

19) Answer: C

Plug in 104 for F and then solve for C.

$C = \frac{5}{9}(F - 32) \Rightarrow C = \frac{5}{9}(104 - 32) \Rightarrow C = \frac{5}{9}(72) = 40$.

20) Answer: C

$5x - 3y = 1$. Plug in the values of x and y from choices provided. Then:

- A. $(-1, 0): 5(-1) - 3(0) = -5 - 0 = -5,$ This is NOT true!
- B. $(3, -1): 5(3) - 3(-1) = 15 + 3 = 18,$ This is NOT true!
- C. $(-1, -2): 5(-1) - 3(-2) = -5 + 6 = 1,$ This is true!
- D. $(2, 1): 5(2) - 3(1) = 10 - 3 = 7,$ This is NOT true!

21) Answer: D

To find the number of possible outfit combinations, multiply number of options for each factor: $4 \times 7 \times 6 = 168$

22) Answer: D

$\text{Probability} = \frac{\text{number of desired outcomes}}{\text{number of total outcomes}} = \frac{13}{21+16+13+15} = \frac{13}{65} = \frac{1}{5}$

23) Answer: A

First, find the sum of six numbers.

$\text{average} = \frac{\text{sum of terms}}{\text{number of terms}} \Rightarrow 42 = \frac{\text{sum of 5 numbers}}{5} \Rightarrow \text{sum of 5 numbers} = 5 \times 42 = 210$

The sum of 5 numbers is 210. If a sixth number that is greater than 48 is added to these numbers, then the sum of 6 numbers must be greater than 210 ($210 + 48 = 258$).

www.mathnotion.com

If the number was 48, then the average of the numbers is:

average = $\frac{\text{sum of terms}}{\text{number of terms}} = \frac{258}{6} = 43$

Since the number is bigger than 48. Then, the average of six numbers must be greater than 43.

Choices A is greater than 43.

24) Answer: A

The ratio of boy to girls is 5:7. Therefore, there are 5 boys out of 12 students. To find the answer, first divide the total number of students by 12, then multiply the result by 5.

$60 \div 12 = 5 \Rightarrow 5 \times 5 = 25$

There are 25 boys and 35 (60 – 25) girls. So, 10 more boys should be enrolled to make the ratio 1:1

25) Answer: B

The perimeter of the trapezoid is 32 cm.

Therefore, the missing side (height) is = $32 - (10 + 5 + 7) = 10$

Area of a trapezoid: A = $\frac{1}{2}$ h ($b_1 + b_2$) = $\frac{1}{2}$ (10) (5 + 7) = 60

26) Answer: B

The probability of choosing hearts is $\frac{13}{52} = \frac{1}{4}$

27) Answer: D

Surface Area of a cylinder = 2πr (r + h),

The radius of the cylinder is 5 (10 ÷ 2) inches and its height is 15 inches. Therefore,

Surface Area of a cylinder = 2π (5) (5 + 15) = 200 π

28) Answer: C

Simplify.

$5x^2y^5(-3xy^2)^3 = 5x^2y^5(-27x^3y^6) = -135x^5y^{11}$

29) Answer: C

The distance between Daniel and Noa is 20 miles. Daniel running at 4.5 miles per hour and Noa is running at the speed of 7 miles per hour. Therefore, every hour the distance is 2.5 miles less. $20 \div 2.5 = 8$.

30) Answer: B

Let x be the number of new shoes the team can purchase. Therefore, the team can purchase $135\,x$.

The team had $35,000 and spent $19,000. Now the team can spend on new shoes $16,000 at most.

Now, write the inequality: $135x + 19,000 \leq 35,000$

31) Answer: D

Let x be the number. Write the equation and solve for x.

60% of $x = 21 \Rightarrow 0.60\,x = 21 \Rightarrow x = 21 \div 0.60 = 35$

32) Answer: B

$\frac{147}{192}$, simplify by 3, then the number is the square root of $\frac{49}{64}$: $\sqrt{\frac{49}{64}} = \frac{7}{8}$

The cube of the number is: $(\frac{7}{8})^3 = \frac{343}{512}$

33) Answer: C

The failing rate is 21 out of 70, $\frac{21}{70}$

Change the fraction to percent:

$\frac{21}{70} \times 100\% = 30\%$

30 percent of students failed. Therefore, 70 percent of students passed the exam.

34) Answer: A

Let x be all expenses, then $\frac{21}{100}x = \$630 \rightarrow x = \frac{100 \times \$630}{21} = \$3,000$

He spent for his rent: $\frac{28}{100} \times \$3,000 = \$840$

35) Answer: C

Isolate and solve for x.

$\frac{5}{9}x + \frac{1}{6} = \frac{1}{2} \Rightarrow \frac{5}{9}x = \frac{1}{2} - \frac{1}{6} = \frac{1}{3} \Rightarrow \frac{5}{9}x = \frac{1}{3}$

Multiply both sides by the reciprocal of the coefficient of x.

$(\frac{9}{5})\frac{5}{9}x = \frac{1}{3}(\frac{9}{5}) \Rightarrow x = \frac{9}{15} = \frac{3}{5}$

TASC Math Prep

36) Answer: C

First, find the number.

Let x be the number. Write the equation and solve for x.

130 % of a number is 104, then:

$1.3 \times x = 104 \Rightarrow x = 104 \div 1.3 = 80$

80 % of 80 is: $0.8 \times 80 = 64$

37) Answer: A

Mrs. Thomson needs an 85% average to pass for four exams. Therefore, the sum of 4 exams must be at lease $4 \times 85 = 340$

The sum of 3 exams is: $72 + 92 + 84 = 248$

The minimum score Mrs. Thomson can earn on her fourth and final test to pass is: $340 - 248 = 92$

38) Answer: C

Let L be the length of the rectangular and W be the with of the rectangular. Then, $L = 7W - 4$

The perimeter of the rectangle is 72 meters. Therefore: $2L + 2W = 72$

$$L + W = 36$$

Replace the value of L from the first equation into the second equation and solve for W:

$$(7W - 4) + W = 36 \to 8W - 4 = 36 \to 8W = 40 \to W = 5$$

The width of the rectangle is 5 meters, and its length is:

$$L = 7W - 4 = 7(5) - 4 = 31$$

The area of the rectangle is: length × width = $31 \times 5 = 155$

39) Answer: B

Use formula of rectangle prism volume.

V = (length) (width) (height) $\Rightarrow$ 3,600 = (24) (15) (height) $\Rightarrow$ height = 3,600 ÷ 360 = 10

40) Answer: D

Use Pythagorean Theorem: $a^2 + b^2 = c^2$

$12^2 + 16^2 = c^2 \Rightarrow 400 = c^2 \Rightarrow c = 20$

TASC Practice Test 2

Section 2- No Calculator

41) Answer: 0.75

Write a proportion and solve for the missing number.

$\frac{80}{12} = \frac{5}{x} \rightarrow 80x = 5 \times 12 = 60$

$80x = 60 \rightarrow x = \frac{60}{80} = 0.75$.

42) Answer: 27

Plug in the value of x and y. $-2(5x + y) + (1 - 5x)^2$ when $x = 1.2$ and $y = -7$

$= -2(5(1.2) + (-7)) + (1 - 5(1.2))^2 = -2(6 - 7) + (1 - 6))^2 = (-2)(-1) + (-5)^2$

$= 2 + 25 = 27$

43) Answer: −4

The equation of a line in slope intercept form is: $y = mx + b$. Solve for y.

$x - 4y = 8 \rightarrow -4y = -x + 8$

Divide both sides by (-4). Then: $y = \frac{1}{4}x - 2$ and the slope of this line is $\frac{1}{4}$.

The product of the slopes of two perpendicular lines is -1. Therefore, the slope of a line that is perpendicular to this line is:

$m_1 \times m_2 = -1 \Rightarrow \frac{1}{4} \times m_2 = -1 \Rightarrow m_2 = \frac{-1}{\frac{1}{4}} = -4$

44) Answer: − 78

Use PEMDAS (order of operation):

$[7 \times (-8) - 42] + [(-3) \times (-14)] \div 7 - (-14) = [-56 - 42] + 42 \div 7 + 14 =$
$-98 + 6 + 14 = -78$.

45) Answer: -72

To solve absolute values equations, write two equations.

$-2x + 1$ can equal positive 17, or negative 17. Therefore,

$-2x + 1 = 17 \Rightarrow -2x = 16 \Rightarrow x = -8$

$-2x + 1 = -17 \Rightarrow -2x = -17 - 1 = -18 \Rightarrow x = 9$

Find the product of solutions: $9 \times -8 = -72$

46) Answer: 123

$-37 = 86 - x$; First, subtract 96 from both sides of the equation. Then:

$-37 - 86 = 86 - 86 - x \to -123 = -x$; Multiply both sides by (-1): $\to x = 123$

47) Answer: 9

$4x - 7 = 3 \to 4x = 3 + 7 = 10 \to x = \frac{10}{4} = 2.5$

Then, $3x + \frac{3}{2} = 3(2.5) + 1.5 = 7.5 + 1.5 = 9$

48) Answer: 512

If the length of the box is 32, then the width of the box is one fourth of it, 8, and the height of the box is 2 (one fourth of the width). The volume of the box is:

$V = lwh = (32)(8)(2) = 512$

49) Answer: 14

Let y be the width of the rectangle. Then; $15 \times y = 210 \to y = \frac{210}{15} = 14$

50) Answer: 60%

Number of males in classroom is: $50 - 20 = 30$

Then, the percentage of males in the classroom is: $\frac{30}{50} \times 100 = 0.6 \times 100 = 60\%$

51) Answer: 42

Let x be the number. Write the equation and solve for x.

$\frac{1}{5} \times 90 = \frac{3}{7}x \to \frac{1 \times 90}{5} = \frac{3x}{7}$, use cross multiplication to solve for x.

$7 \times 90 = 3x \times 5 \Rightarrow 630 = 15x \Rightarrow x = 42$

52) Answer: 53.25

$Average = \frac{sum\ of\ terms}{number\ of\ terms}$

The sum of the weight of all girls is: $25 \times 35 - 875$ kg

The sum of the weight of all boys is: $5 \times 38 = 190$ kg

The sum of the weight of all students is: $875 + 190 = 1,065$ kg

$Average = \frac{1,065}{20} = 53.25$

"End"

www.ingramcontent.com/pod-product-compliance
Lightning Source LLC
Chambersburg PA
CBHW080510090426
42734CB00015B/3017